Economic Analysis of Fermentation Processes

Harold B. Reisman, Ph.D.

Director of Manufacturing
Formulated Foods Group
Stauffer Chemical Company
Westport, Connecticut

CRC Press
Taylor & Francis Group
Boca Raton London New York

CRC Press is an imprint of the
Taylor & Francis Group, an **informa** business

CRC Press
Taylor & Francis Group
6000 Broken Sound Parkway NW, Suite 300
Boca Raton, FL 33487-2742

Reissued 2019 by CRC Press

© 1988 by Taylor & Francis Group, LLC
CRC Press is an imprint of Taylor & Francis Group, an Informa business

No claim to original U.S. Government works

A Library of Congress record exists under LC control number:

Publisher's Note
The publisher has gone to great lengths to ensure the quality of this reprint but points out that some imperfections in the original copies may be apparent.

Disclaimer
The publisher has made every effort to trace copyright holders and welcomes correspondence from those they have been unable to contact.

ISBN 13: 978-0-367-20626-0 (hbk)
ISBN 13: 978-0-367-20628-4 (pbk)
ISBN 13: 978-0-429-26256-2 (ebk)

Visit the Taylor & Francis Web site at http://www.taylorandfrancis.com and the
CRC Press Web site at http://www.crcpress.com

INTRODUCTION

The last decade has been marked by enormous change in the life sciences. New bioactive entities have been made and novel techniques have opened new vistas in molecular biology. Monoclonal antibodies offer a reasonable hope for targeted drug missiles to cure disease; many new diagnostic methods are based on this advance. New vaccines are being prepared from specific protein moieties offering disease prevention without side effects. In microbiology, classic mutation methods have been superceded by protoplast fusion and recombinant DNA technology. A host of new compounds arises monthly and new processes are presented which offer potential for mass production of well-known biodynamic molecules as well as the ever-growing list of newer ones.

The more mundane area of economics (as applied to what is called, for simplicity, biotechnology) has only recently received more serious attention in the business community. Sooner or later, one must move from concept, discovery, or laboratory preparation to sale of a desired material. This simple fact is true regardless of the material's nature or its derivation. The many new biotechnology companies are finding that issuance of stock and even a patentable discovery are not sufficient to maintain long-term corporate viability. Many of these companies may find themselves in the position of creators, holders, or purveyors of technology while they are, or may become, totally dependent upon larger competitors (major pharmaceutical companies) to do actual production, marketing, and sale of resultant products.

It is the purpose of this book to outline and detail the many steps which are involved in bringing a fermentation product to market. Ultimately, investment must result in a monetary return (unless there is some other overarching goal). Many of the steps are applicable to the production of vaccines, antibodies, bioactive peptides, and so forth, but the basic orientation is that of a fermentation product. No single text can cover in depth all necessary planning, scheduling, construction, costing and marketing operations that must occur; however, there is enough detail given so that anyone with a reasonable technical background will be aware both of the actual steps needed and the methodology used to complete each step effectively and efficiently (see Figure 1). Finally, return on investment and sensitivity analyses are reviewed to bring the economic picture into focus.

The potential of the microorganism is legion. A few examples suffice. Solvents and precursors (acetone, butanol) can be made by anaerobic fermentation. Microbial polysaccharides can be used as food additives and in enhanced oil recovery. The potential for food, feed, and ethanol production from waste cellulosic materials via microbes exists. Textured mycelial food products that simulate veal and chicken have been produced. Psychoactive and immunoactive compounds can be produced microbially. Microbial synthesis of interferons, rennet, and growth promoters is now possible. It must be noted that only some of the above processes are, or will be, commercial. Some may move to commercial scale in the future. The major determinant, at least in the West, will not be technical feasibility, but economics.

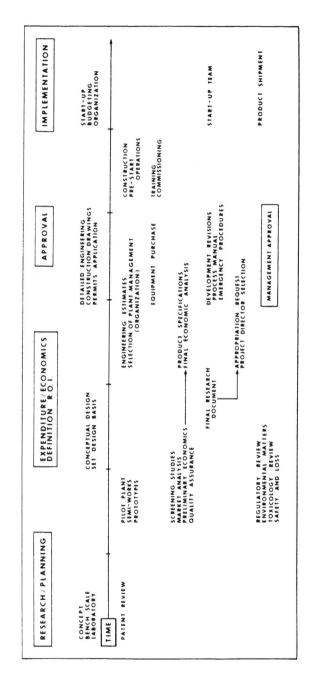

FIGURE 1.   The project cycle.

THE AUTHOR

Harold B. Reisman is Director of Manufacturing for the Formulated Foods Group of Stauffer Chemical Company. He has over 25 years of industrial experience in fermentation design, product isolation, scale-up, pilot plant operations, fermentation plant construction, start-up, and operations. Products include antibiotics, vitamins (riboflavin and B_{12}), amino acids (lysine, monosodium glutamate), therapeutic enzymes, interferon inducers, as well as novel food ingredients. Current responsibilities include manufacturing operations, research, development and applications support for the Food Group, quality assurance and control, and regulatory and environmental affairs.

He received his undergraduate degree in chemical engineering from Columbia University, New York, and a Master's degree working with Prof. R. K. Finn at Cornell University, Ithaca, N. Y. He returned to Columbia for his doctorate in chemical engineering working with Prof. E. Gaden on optimization and kinetic analysis of penicillin fermentation. A Fulbright Grant supported a year's postdoctoral research at the Istituto Superiore di Sanita in Rome working in a group headed by Prof. E. Chain. Dr. Reisman was then employed as Section Manager in the Merck, Sharp and Dohme Research Laboratories in biotechnology and fermentation development. A number of papers were published and a patent on continuous fermentation ensued.

He joined Stauffer in 1973 and was Bioengineering Laboratory Director, Plant Manager, and Business Group Director prior to assuming his current position. He has visited Japan more than half a dozen times and has been involved in technical and commercial exchanges with three major biotechnology firms there.

He is a member of the American Institute of Chemical Engineers, the American Chemical Society, and the Institute of Food Technologists. He has authored chapters in *Microbial Technology* and the *Encyclopedia of Chemical Processing and Design*.

ACKNOWLEDGMENT

I would like to thank my wife, Miriam F. Reisman, for producing the excellent artwork contained within these pages.

For Miriam, Jocelyn, and Joseph

TABLE OF CONTENTS

Chapter 1

RESEARCH AND DEVELOPMENT

I. RESEARCH ORGANIZATION

The research organization is a critical determinant in the economic success of most companies. This is all the more so in a high technology business. Successful development of new biological products depends upon integration of the research organization at all levels in the company. The person in charge of the research organization should be included in the highest policy-making level of the company. Research interfacing with all staff groups, manufacturing, and especially marketing is imperative to smooth and accelerate product introduction.

Characteristics of a smoothly functioning and productive organization are

1. Responsive to changes in market need (including means of delivery); marketing and sales information flows constantly to research managers; program corrections are made so that solutions match real problems.
2. Selectivity is exerted so that available resources (manpower, money, materials) are not dissipated, but applied and focused on strategic programs.
3. There is an ongoing search for novel, but related product concepts, while reexamination leads to dissolution of programs which show little promise or progress.
4. Functional levels are kept to a minimum. There is movement of people to match needs whether they arise in the research laboratory, the pilot plant, the operating plant, or the marketplace. Structure is fluid.
5. Innovation is fostered and rewarded. There is a climate of creativity that is recognized within and without the company.
6. Research results are reported (not overreported) in a timely fashion. A sense of enthusiasm pervades the research group and members are anxious to resolve issues and let others know of that resolution. Progress is made (translation to production scale, for example) without all possible design data in hand. Risks are mutually understood and accepted.
7. Corporate objectives are clear and there is an agreed-upon balance between short- and long-term projects. Strategic decisions are made or changed with intimate involvement of key research personnel.
8. Research personnel have, or are taught, a financial understanding of the company (and industry) and so can understand and explain business implications of various modes of action.

Why spend this effort on research and development when the subject is "economics"? The answer is clear. The commercial success of a venture or a fermentation company is tied to research success; furthermore, the cost of research is very high and going higher. The difference between success and failure can easily be measured in tens of millions of dollars; it is not unusual to involve swings in the hundreds of millions. Clearly, the costs of numerous failures will mean shrinkage or dissolution of the company. Some 15 years ago, cost per professional research staff member was $50,000 annually. Even at an inflation rate of 5%, the annual cost now would be slightly over $100,000/year. Indeed, published information supports the rough estimate.[1] A survey of 157 industrial research organizations indicates that operating cost per professional was $119,000 in 1985 compared to $113,000 in 1984. Fifteen pharmaceutical companies were included in the survey; for that set, cost per professional was $84,000 in 1984 and $91,000 in 1985. For the pharmaceutical group, R & D expenses as a percent of sales averaged 7.5 in 1984 and 7.3 in 1985.

Data for four large companies (full year 1984) are shown below:[2]

	Sales ($million)	Earnings ($million)	R & D expenditures ($million)	R & D as % of sales
Eli Lilly	3109	490.2	341	11.0
Pfizer	3855	507.9	252	6.5
Merck	3560	493.0	393	11.0
G.D. Searle	1246	161.6	120	9.6

Earnings exclude extraordinary and nonrecurring items. R & D expenditures are a very high percentage of net income; values of 50 to 80% are common. For start-up firms, R & D costs may be many times net earnings, if there are earnings at all. For 30 pharmaceutical firms, R & D expense was 6.7% of sales (on the average) and accounted for 40% of pretax income (also an average value).[3] The average R & D cost per company employee was $5704. To compare to other research-oriented groups, average R & D expense as a fraction of sales for the chemical industry was 3.0% and for the electronics industry, 4.3%. Even with these levels of R & D expenditures (some might conclude *because* of these levels of expenditure), Lilly, Merck, and Pfizer can be found among the top 100 U.S. firms when ranked in order of corporate cash flow.[4] Their respective returns on equity were 22.1, 19.4, and 20.7%; these are very respectable figures.

There are about 400 firms that can be called biotechnology oriented. It is a matter of conjecture as to how much of the orientation is real and how much is capitalization on an area of great interest. Among the so-called start-up companies, R & D expenditures (1984) range from less than $1 million (Ribi) to about $55 million (Genentech). What is consistent is the high ratio of R & D expense to total expense. Table 1 is a compilation of data for selected biotech companies of differing sizes. Expenditures for R & D are often equal to, or greater than, annual sales figures.

A ten-person team (with necessary support structure) is not very large even for a start-up company. Cost is in excess of a $1 million/year. Ten man-years (or in more conventional terminology 120 man-months) pass very rapidly; most often, multiyear commitments are essential. The reason for inclusion of this subject is clear, as is the need for careful research planning.

While much of the interest in biotechnology is focused on the smaller, start-up firms, major changes have occurred in very large organizations as well. In the early 1970s, Monsanto spent less than 3% of sales on R & D. The figure has moved to greater than 5% of sales a decade later, with dollar spending on R & D approaching 400 million. Much of this transformation is related to a shift to biotechnology, agricultural chemicals, and health care. This is a major restructuring of a very large chemical company. Many aspects of the strategic change and the company's research and development structure are detailed in a very useful article on Monsanto, including an interview with H. A. Schneiderman (senior vice-president of R & D).[5] In 1985, Monsanto purchased G. D. Searle, thus, making a further and major commitment to pharmaceuticals and biotechnology. The cost to Monsanto was $2.7 billion. Not only is a massive change for a corporation involved, but there is a directed effort into new business and technical areas. Cultural changes are involved, a new regulatory outlook has been introduced, and a new time frame for development is in place. Technical-marketing interactions and needs are explained. It will be instructive to read this article and monitor the company in the next decade. The picture that emerges after expenditure of hundreds of

3

Table 1
R & D EXPENDITURES AND SALES FOR SELECTED START-UP COMPANIES: 1984[a]

	Revenues		Expenses		Net gain or (loss)
	Sales	Interest	R & D	Total	
Amgen	2.78	3.33	8.76	11.06	(4.94)
California Biotech	6.71	1.56	7.26	8.34	(0.13)
Centocor	10.88[b]	1.96	6.72	12.14	0.70
Cetus	35.85[b]	10.37[c]	31.41	45.16	0.99
Damon Biotech	2.39	2.66	3.82	8.55	(3.51)
Genentech	65.63[b]	4.16	54.98	66.78	2.72
Hybritech	14.60 16.23[d]	3.11	13.70	32.66	(1.83)
Molecular Genetics	6.37[b]	2.85	4.81	9.86	(0.64)
Monoclonal Antibodies	1.62[b]	0.59	1.39	5.83	(3.62)
Ribi Immunochem	0.52	0.40	0.43	1.20	(0.29)

[a] Dollars in millions.
[b] Includes "sponsored research" or "contract revenue".
[c] Includes "other income" of $0.73 MM.
[d] Listed as "contract revenues"; total operating revenue $30.83 MM.

millions of dollars in capital and equal or greater sums in R & D expense will present a classic case study whatever the economic outcome.

Setting of product objectives is a management decision. It is imperative that once these objectives are communicated, two important summaries be detailed and recorded. One concerns allocation or development of capabilities and resources. The second involves total resource allocation, including funds needed for research and development, regulatory affairs and clearance, capital (or lease/rental), and marketing. Both summaries must be updated on a routine basis. It is obvious that the early summaries are merely best estimates and may have a large margin of error. It should be equally obvious that even 3 months of lab work, legal and regulatory review, and marketing analysis will stimulate major revisions in the "first pass" summaries. The changing resource needs and resource (cash) flows must be updated and communicated to those responsible for setting the product or process objectives. While the first steps — resource allocation and flow of funds — are often taken, due regard to follow-up may be lacking. The seeds of discord or failure are often sown as revision is cursory or disregarded altogether.

The *planning* phase (Table 2) involves, first, defining and communicating the objective. In order to achieve the objective, a list of activities is required. Not only are certain activities to be done, they must be completed in a logical sequence in an approved time frame. The establishment of a logic sequence in time may involve various bar charts, PERT charts, networking, scheduling, or computer-generated sequences. The planning sequence may involve none of these, but some sequence (even if a mental image) must be mutually agreed upon and should be followed. However, the hazards of following an abstract mental image should be understood.

Once the planning phase is complete, the *allocation* procedure must be followed. New and available resources are compiled. The work schedule is set and budgeting requirements are detailed. All overheads are included whether by factor or by line account. Necessary interfacing with legal and regulatory personnel (in-house or out) should be programmed. If

Table 2
THE PLANNING SEQUENCE

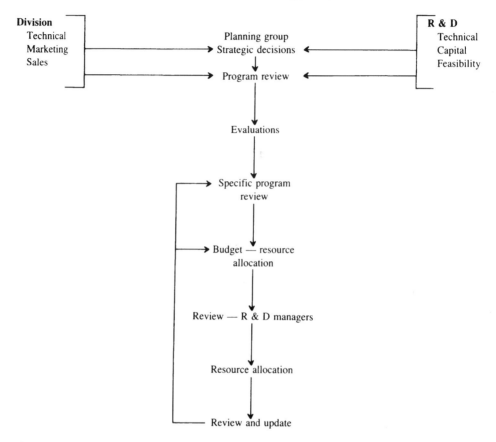

After *Innovation*, No. 19, 51, 1971.

the product is to be marketed at a known future time, critical points in the logic sequence must be established. Clearances are a fact of life and the sooner the requirements and their fulfillment are programmed in, the better the hope for project completion. A cash flow forecast should be included in the allocation phase. While not absolutely essential at this point, certain problems that seem likely can be identified and contingency plans made. In the subsequent *update* phases (which will probably result in plan modification, timing changes, and reallocation of resources), it is essential that problems be identified and multiple contingency plans be prepared.

The pharmaceutical industry is highly innovative, but it must be considered market driven. A commercial enterprise would not seek a specific antibiotic against an ubiquitous and harmless microorganism unless there were some ulterior motivation; similarly, for a monoclonal antibody against a circulating protein that signaled nothing in a physiologic sense. In general (and as compared to a "heavy" industry), the industry is characterized by relatively modest capital investment relative to the value of the product. Investment is high relative to quantitites produced. The industry norm is to have a multiplicity of products and a reasonable new product flow to compensate for product obsolescence. The term "obsolete" refers more properly to displacement by more active materials having fewer side effects. The objectives of research in the fermentation industry are

1.   Screening for, and selection of, novel bioactive moieties and the means for their synthesis and purification

2. Improved functional performance of a precursor, a novel compound, or a derivatized natural or synthesized product
3. Cost reduction for novel or existing product formation by applying microbial, chemical, biochemical, and engineering techniques
4. Scale-up and commercialization of any and all of the above

A short summary of the many steps needed to bring a discovery to mass production has been published; product requirements for clinical or field trials are given.[6] Interrelationships between strain improvement, medium development, and process optimization are shown. Examples of successful development are described.

Understanding the value of R & D in a qualitative way is simple when a number of new products have been developed in a short time interval. Putting this understanding in a quantitative form is far more difficult. Even when accomplished, heated discussion often ensues. Normal accounting indicators, such as profit from operations (PFO), return on investment (ROI), and profit-to-sales ratio, are not adequate if applied directly and indiscriminately to annual research expenditures. Some assumptions and weighing are needed to quantify research productivity.

The quantitative evaluation of research and development is essential (if only for proper research planning) and it is possible. Both tangible and intangible factors can be listed and weighted. Not least of these factors is how much a product or process contributed to corporate goals by being translated into dollars. All aspects of research (fundamental, applied, engineering, start-up, competitive evaluation) can be judged by how much has been generated in dollars by year over a reasonable time — compared to the expenditure. The simplest format is to first divide the corporation into manageable "categories". (For a start-up company, the number may be one or two.) Each business category should have a percentage of the total research effort "assigned" to it. Ideally, a 5-year history can be used. The successful projects (criteria must be established for "success" — monetary return is not a bad starting point) should be listed. The gross profit contributed by these successful projects can be summed. Once again, a sufficiently long history is needed. Five years seems like a minimal period. A very coarse ROI can be calculated by summing the profit contribution for each business and dividing by the cost of R & D in that same business. It may be necessary to project future earnings in some cases. One can, of course, use this simple model in the planning cycle, that is, determine a minimum acceptable return on research investment and fund the necessary projects so that a reasonable success rate gives the desired return. It is rather more complex than given here, since the number of variables and degree of uncertainty are very high at the initiation of a program.

Decision analysis has been used in evaluation of R & D projects. Thomas[7] discussed strategic management of R & D in two detailed cases — one in the electronics industry and one in the ethical pharmaceutical industry. He notes that in recent years pharmaceutical companies have limited research areas, so that expertise and an understanding of the market can be built up leading to success in closely related, integrated product areas. One not insignificant reason for focusing is the fact that new drug development and clearance might involve "investments in excess of $50 to 75 million". The specific study involves licensing and a joint venture strategy with significant up-front investment. Decision analysis with decision trees (toxicological trials and clinical trials with different scenarios) is reviewed. Risks were rated as was reward (present values analyses at high to low probability). The findings, which seem to apply to most research management situations, can be detailed under the headings of assessment problems, discounting processes, flexible decision criteria, and the process of policy dialogue. A few of the conclusions are

1. Balancing risk preference and time preference is very important in R & D management.

2.   Managers want multiple performance measures for portfolios and individual projects.
3.   The role of decision analysis was to provide output as a function of numerous variables and express results in terms of dynamic interactions. One criterion of utility was not adequate.
4.   Profiles of cost, revenue, cash flow, and capacity encourage awareness of risk and uncertainty on an individual project and a portfolio of projects.

Schmitt presents an overview of corporate R & D and, also, discusses overall corporate R & D strategy.[8] The key points of emphasis are similar to those noted above — synergy, interdisciplinary focus, and lead time. A pragmatic or a "trying" approach between the technology and marketing sectors is needed to progress at a sufficiently rapid pace. Once more the point is made that ROI does not tell it all.

An important survey and analysis of research and development productivity is available in a two-part series.[9] It is possible to construct a framework which not only identifies key productivity factors, but also selects methods which will measure and improve economic return. Out of the many possible activities, 13 were identified and ranked by many research directors as having the greatest impact on R & D return. These high return activities are listed below, in order of importance, and many suggestions are given that relate to these efforts.

Rank
1   Identify customer needs
2   Professional personnel quality
3   Coupling to technical efforts — marketing
4   Identifying projects
5   Identifying technical possibilities
6   Demand outlook
7   Project staffing
8   Strategies of competitors
9   Coupling to technical efforts — manufacturing
10   Project planning
11   Identifying limits
12   Project termination
13   Characterizing technology

The R & D yield is defined as the profit made from improvement in technical performance. The first and third rankings relate to "yield" in that both must be successful to maximize economic return.

There are certain other major interrelationship problems that must be overcome. These refer to research interactions with other parts of the company. Even if overcome in one time period or in one project, these problems seem to reappear in a cyclic fashion. The first concerns monitoring of a project. Not only does research management have an interest in focused effort to achieve timely results, there are other corporate functions — manufacturing and marketing are merely two obvious ones — that are deeply involved in research expenditures, process details, product configuration, and, most of all, timeliness. The subject of how to keep projects on track is discussed in a clear and organized fashion by Szakonyi.[10] (He discusses four general "needs" or "processes" that must be satisfied whether the project reviewer is within the research organization or elsewhere in the company.)

1.   Tangible results. Some progress must be visible as time passes. Milestones must be met. Stepwise progress must be obvious even to the untrained eye. The sense of what

might be called "partial accomplishment" will be an ongoing positive reinforcement of research personnel and those in other groups.

2. Compare technical progress with costs. It is probably true that given unlimited funds and unlimited time, any goal can be reached. Given constraints of money and time, it is imperative that progress must be related to expenditure. Costs must be viewed retrospectively and prospectively. If so much has been spent to get to 30% of where we thought we would be, how much more will be realistically needed to complete phase I of the project? Hard decisions must be made when costs incurred and progress attained show serious discrepancies.

3. System of progress reports. While voluminous or tedious documentation is invariably a waste of time, no documentation will result in the same, or greater, waste of time and money. Generation of paper is to be avoided, but a routine and periodic release of progress reports is necessary. This is a good idea even if there is nothing positive to report; some might add that in such a case, it is even more important that a report be issued. All interested personnel, whether line or staff, must be informed on status of effort.

Szakonyi[10] has a terse human relations summary on research interactions with others in the company:

"Rather they will be based on conditional agreements that both sides will continue to cooperate as long as they both continue to gain something from their cooperation. The trick to keeping R & D projects on track, therefore, involves keeping these conditional agreements viable by making sure that the objectives of the R & D project coincide as closely as possible to the diverse and specific objectives of the people involved."

The second major interaction problem (or opportunity) occurs on translation of a process to manufacturing. There are a host of factors (mainly economic) which come into play and this situation will be discussed in a subsequent section on start-up.

II. PROCESS DEVELOPMENT AND SCALE-UP INFORMATION

The aim of process development is quantitative delineation of all key factors which will permit timely and economic design and construction of an operating plant (or part of a plant) which will then perform in a stable, predictable, and cost-effective manner to produce a desired product of known quality at a determined rate. There are many factors that will impact cost and each must be considered. The degree to which quantitative definition (involving prediction equations, mean values, standard deviation, sensitivity analyses) is needed depends upon the specific process and product, and the very determination of emphasis sets the tone for later success or failure. Great precision and detail covering a weak interaction or tertiary variable not only does little to aid process design or operation, but the effort detracts from an understanding of primary variables.

Key factors in process development include:

Raw materials	Material balance
Strain selection	Component balance
Strain maintenance	Type/extent of recycle
Media development	Energy requirements
Process optima	Aeration-agitation needs
Waste streams	Sensitivity to upsets
Quality control	Downstream recovery
Equipment design	Stability (in-process, product)

A simple outline should be prepared so that available laboratory or published information is listed and unknowns are highlighted. Modifications can be readily made as development proceeds. Such a simple outline is given below for a well-known fermentation, that for citric acid:

Carbohydrate	Beet molasses
	Glucose syrups
Nitrogen source	No other additive if molasses used/salts
Process	Submerged or tray
Efficiency	
Fermentation	90 + %
Isolation	90%
Microorganism	*Aspergillus niger* (with identification)
Sugar, initial	15—18%
Cycle time	
Submerged	3—5 days
Tray	10 days
Turnaround	10 hr
Temperature	25—30°C (may be staged)
pH	2.5—3.5 (controlled)
Airflow	0.5—1.0 vol/vol/min
Pressure	1.5 atm
By-products	Mycelium, oxalic acid
Pretreatment/additives	Methanol (2% to fermentation)
	Cation/ash levels critical
Isolation	Settle/filter mycelium
	Precipitate as calcium salt
	Redissolve with H_2SO_4
	Demineralize/decolorize
	Crystallize and dry

Reviews on process design and scale-up have been written. A review by Lilly[11] is useful, especially since he notes that one " . . . may feel that it is a wonder that anything has been scaled up". A history of fermentation development and background information on fermentation design and operations is given by Bjurstrom.[12] There are a number of books which give a good theoretical background to various aspects of development and scale-up while containing many examples involving industry practice. An academically oriented text by Bailey and Ollis[13] is worth consulting. Transport phenomena are well described and design examples (sterilization, reactors) are given. Blakebrough[14] edited two volumes that combine theory and practice. The second volume contains chapters that cover some aspects of downstream processing (such as evaporation, drying, heating, cooling, use of radiation). There is some extension into the foods area, but many examples of industrial equipment are given. Perlman[15] discusses products and producers (commercialized through 1977) and covers raw materials, fermentor design, and an overview of recovery.

During laboratory and pilot plant development, alterations in key process variables will have been studied. Some of the variables were purposefully changed in optimization studies and some resulted from process upset, mischarge of nutrients, or "unknown" factors. The overall picture that emerges should point to those parameters which must be controlled within a narrow band and those variables that might be permitted to range somewhat more widely. The next step concerns creation of a micro- and macroenvironment, on a large scale, that will allow maximal rate and yield to be obtained *within economic constraints*. The last point is sometimes neglected at costs which become apparent at a later time.

Some simple examples will suffice for now:

1. Very often, use of any reasonably effective defoamer depresses yield by 5 to 10% regardless of other changes. Scale-up fixes operating volume at 65% of nominal production fermentor volume to ensure no foam loss under highly aerated conditions. However, if defoamer use is "permitted" and a 7 1/2% yield loss assumed, operating volume could be 75% of nominal volume and overall recovery would be improved by 6.7%. Even a 10% yield loss means enhanced overall product formation with use of defoamer.

2. Optimum productivity is achieved at a specific oxygen transfer rate (OTR) and dissolved oxygen (DO_2) level. At the desired operating volume, a specific power input and turnover time is required to satisfy the OTR and DO_2 constraints. The drive, motor, seal, and shaft assembly require a step change in duty/diameter/mechanical support. No balance is made between the major increment in capital investment (plus maintenance and on-going energy cost) and the modest yield or rate reduction corresponding to somewhat less stringent agitation-aeration requirements.

3. A pilot extraction operation involves multistage solvent extraction and carbon purification. Recoveries of solvent are 98% and product recovery is 96%. These extraction stages are scaled to a production facility. No calculation is made of recovery and total operating cost (including changes in labor, materials, maintenance, energy, depreciation, taxes, insurance) if solvent recovery is only 96% and product recovery is 93%. Just as in heat recovery, the maximum in recovery most often does not correspond to the economic optimum or least cost point.

While there is no intent to give an in-depth review of critical parameters, certain points will be covered that are of major significance. Reviews are available for more detail and analysis.

III. MIXING AND OXYGEN TRANSFER

The first and foremost scale-up parameter in fermentation design is mixing. It is rather remarkable that the earliest fermentors for antibiotic production (constructed in the late 1940s and early 1950s) remain usable today for many new generations of microbial product. The use of a stirred tank gas-liquid contactor has endured for four decades; there is no indication that the design will soon be retired. In the interim, there have been reports on aerated columns (tower fermenters), oxygen supplemented fermentation, draft tube vessels, disk or rod agitators, vibrating mixers, external circulation systems with pumps, and gas lift fermenters. At best, some of these designs have found selected niches; others are laboratory curiosities. Clearly, the sparged agitated contactor performs in a manner that satisfies physiologic requirements of many microorganisms.

Not only must a unit volume be satisfied (temperature, dissolved oxygen, nutrients, pH, redox potential, pCO_2), but vessel turnover must maintain quasi-homogeneity. In a shake flask or laboratory fermenter, turnover time is minimal; it can be measured in seconds. Gradients are minor or nonexistent. Any material added — whether it is substrate, acid, or base — is mixed into the bulk of the fluid with negligible delay; that is, every unit volume "sees" predetermined optima with little or no delay. At most, some seconds pass before recirculation occurs. For oxygen transfer, the critical region is the impeller zone where initial shear on the gas bubble occurs. This is the region of maximum DO_2; recirculation time to this region determines whether or not oxygen deprivation occurs.

A simple calculation shows the potential for a problem:

	Case 1	Case 2
	Case 1	**Case 2**
Oxygen transfer rate (OTR), mM/ℓ min	0.417	1.667
Saturation, mM O_2/ℓ	0.2	0.2
Seconds to zero DO_2	28.8	7.2

In a laboratory fermentor where mixing time is 5 sec or less, both fermentations will show no problem as relates to oxygen supply and demand; when mixing times reach 15 to 30 sec, one can anticipate a different response to oxygen deprivation. While oxygen deprivation is of major importance, many fermentation additives are toxic or inhibitory if not present below certain concentrations. Continuous, low-level addition is the remedy used; however, if mixing is inadequate, gradients develop with resulting negative responses. Solids in broth or elevated viscosity will magnify the untoward response and provision for adequate turnover must be provided. Lastly, maintenance of a high heat transfer coefficient (to jacket or coils or both) in a reduced density gas-liquid broth requires vigorous mixing. Temperature gradients are to be avoided as cell growth or product formation normally shows rather discrete optima.

Although data have been published on mixing time, most companies supplying agitation systems are loathe to present performance information. Some generalizations are possible which will give expected ranges and direction for scale-up. Reviews on aeration and agitation have been published over the years. Blanch details various methods of predicting and correlating power input to gas-liquid systems.[16] A number of novel reactors are also described. Yoshida[17] presents a theoretical background including measurements of interfacial area. Effects of electrolytes, surfactants, and system rheology are discussed. Ratios of power draw in gassed/ungassed systems are given and some scale-up data are presented. There are 115 references listed. Humphrey[18] discusses agitation-aeration scale-up as well as general principles of biochemical engineering.

Satisfaction of the oxygen demand of a culture is always possible, but both capital and operating cost might be prohibitive, especially on scale-up. The oxygen demand or maximum oxygen uptake rate is dependent upon substrate, microbe, and process parameters such as temperature and pH. Uptake or demand, as so defined, cannot be limited by physical inputs such as power per unit volume or oxygen supply. The actual OTR varies in the course of the fermentation, indicating changing respiratory requirements. For oxygen transfer, as for other key variables, maintenance of a fixed input (or a single point variable for temperature, pH, or pressure) cannot offer a continuous optimum for an evolving, dynamic system. Continuous or step changes in input variables might be conducive to improved performance. If aeration and agitation can be reduced as the fermentation proceeds, output might be enhanced while important energy savings are realized. The same is obviously true for temperature and other variables.

There is a well-known plot whose general form is applicable to many minimum cost determinations. The plot may be used to determine minimum cost — both capital and operating — of supplying necessary oxygen. In Figure 2, there are two intersecting plots. One relates dollar cost (either capital or operating) for air compression, that when added to an agitator power input (along a vertical axis), will satisfy a specific oxygen demand. That is, a very high aeration power input plus a relatively low agitator power input (left section of plot) will give a combined power input to satisfy a preselected or predetermined oxygen demand. Parameters fixed for the plot are that of scale, oxygen transfer capability, and media (process). Each independent power input has a cost associated with it (ordinate). Capital cost should include required substations, starters, buildings, and installation. It is

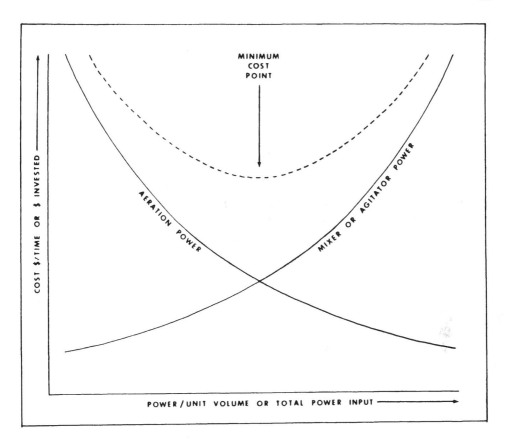

FIGURE 2.   Cost vs. power input (agitation and aeration).

clear that a nonlinear relationship exists between capital cost and installed mixer horsepower; structural, support, and seal cost requirements will rise exponentially once a certain size is exceeded. Operating costs should include direct utility costs, taxes, insurance, maintenance, and depreciation, although a lumped percentage figure related to investment is possible. A minimum cost point (or perhaps better, a minimum cost range) exists that will not only give a true economic minimum, but will also satisfy other scale-up requirements. For example, at too high a superficial velocity, foaming may be excessive. At too low an agitator power input, turnover time may be excessive and regions of zero DO_2 will exist. Evaluation of pCO_2 must also be considered as certain fermentation processes are adversely affected by high exit CO_2 concentrations.

Ultimately, the effect of design parameters on OTR must be determined or estimated from published correlations. The standard form of the equation used to determine oxygen transfer is

$$OTR = K_L a(C^* - C)$$

where OTR = oxygen transfer rate (units may be mM/ℓ hr), K_L = oxygen transfer coefficient, a = interfacial area, C^* = equilibrium dissolved O_2 concentration, and C = actual dissolved O_2 concentration. $K_L a$ is usually given as a combined parameter (units of reciprocal time) to obviate the need for interfacial area determination. In well-mixed vessels, C may be taken as the value in equilibrium with O_2 in the exit gas. This is generally true for smaller fermentors. For larger or production scale vessels, a log-mean average concentration, based upon inlet and exit gas concentrations, can be used. OTR is dependent upon power input, impeller design and spacing, aeration rate, pressure, temperature, broth characteristics (surface tension, viscosity, particulates), impurity levels (foam and defoamers especially), and gas holdup. One empirical correlation that may prove useful if certain parameters are more or less constant is given by Humphrey.[18]

$$K_L a = K^1 (P/V)^\alpha (V_s)^\beta$$

where $K_L a$ = volumetric oxygen transfer coefficient, K^1 = equipment (correlation) constant, P/V = power input per unit volume, and V_s = superficial air velocity. For pilot scale, $\alpha = 0.95$ and $\beta = 0.67$. At larger scale, $\alpha = 0.4$ and $\beta = 0.5$. The correlation of interest is that between cell growth rate or specific product formation rate as a function of specific oxygen uptake. Once yield data vs. oxygen uptake is known, scale-up to give desired formation rates is undertaken. In very general terms, OTR falls between 50 and 175 mM/ℓ hr and power inputs are found between 1 and 2 hp/100 gal for production fermentors. Both higher and lower values have been reported and variations are due to substrates used and microbial physiology for the specific case. A review of many different procedures for measurement of OTR is available.[19] A brief analysis and critique of each system are given. Resistances to mass transfer are discussed separately. A review of the various models and critical parameters is included. The importance of residence time distribution and hydrostatic head is noted. Some newer developments offer greater precision in estimating aeration capacity.[20] A computer-controlled fermentor was used to determine time delays, impact of noise, and other variables; ultimately, an algorithm was developed to correct for various potential error functions. In another study, microbial product distribution was shown to be related to local DO_2 concentration.[21] The authors utilized acetoin and butanediol formation as an oxygen-sensitive biological system which can be used to study mixing and oxygen transport. The system described can be used to detect dead zones and oxygen gradients. As such, it could be a powerful scale-up tool.

IV. HEAT TRANSFER

There are two major heat transfer situations in fermentation. (Extraction and concentration steps and heat transfer requirements depend upon the process employed and hence, are more specific.) One concerns sterilization of vessel and media (including additives) and the other concerns heat removal due to microbial metabolism.

A detailed section on media sterilization is given in a text.[22] Continuous sterilization is a clear choice for media preparation, whereas nonthermal sterilization may be required for mammalial cell culture or other fastidious organisms (Figure 3). Advantages of continuous sterilization are

1.   Improved equipment utilization (shorter turnaround)
2.   Amenable to automation and computer control
3.   Averaging of utilities demand
4.   Minimizes thermal damage to media
5.   Reduced contamination frequency
6.   Simplified cleaning of fermentors

The pressure in the retention section (and elsewhere on the sterile side) should be sufficiently high to prevent flashing. Once some cooling occurs, as in the regenerative heat exchanger, flashing is less of a problem, but should always be considered and avoided. Any flashing will negatively impact the residence time distribution and has potential for a break

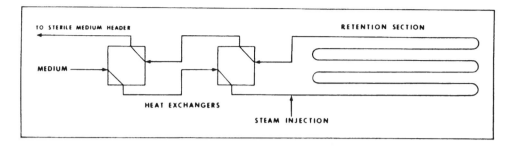

FIGURE 3. Configuration of continuous sterilizer.

in sterility. Pressure indication and a pressure control valve are essential elements of control as are temperature indicators, one of which will control steam input. Direct steam injection is common and dilution must be included in volume determination.

The microbial heat generation rate is defined and derived from yield factors or heats of combustion data.[13] The simplest formulation is one in which only cell mass is produced; other metabolic pathways require similar calculations based upon products formed. Very often, the rate of heat evolution is directly proportional to the rate of oxygen consumption. This is not a surprising conclusion if the predominant oxidizing agent is oxygen; this is normally the case. During pilot fermentations it is useful to determine cooling loads (by flow measurement and temperature differentials) even though radiation loss or gain is high. Substrate utilization rates and yield values can be used to determine correlations. Large variations should be investigated further.

The costs incurred in transfer of heat are considerable. Whereas mass transfer considerations receive a great deal of attention, a similar effort is needed for heat evolution and heat removal. It is rare that a jacket alone is sufficient for temperature control. Such cases should not be overlooked, however, because of reduced cost. Whereas the flexibility of such a fermentor is limited (alternate culture or alternate substrate may mean a higher heat load), cost savings on installation are appreciable. Furthermore, cleaning is simplified. Use of internal coils, while often essential, should be justified in each case. While metabolism of the organism contributes the major heat load, there are other inputs that must be considered. Heat input due to dissipation of agitator power may be an important input. Sensible heat from all input streams should be calculated. Heat gain or loss from the environment is usually not a critical factor, but outdoor location or wide-ranging ambient temperatures might cause control problems.

External heat exchange for fermentor broth has been suggested and some designs have been tested and run at large scale. Sterility considerations are obviously important in such a recycle system. It is also true that heat exchange surfaces must be cleaned on a fairly routine basis. Heat exchangers used in sterilization of media, water, and fermentation additives must be constructed of corrosion-resistant material, must be sealed or gasketed so there is no possibility for entry of foreign organisms, and must be designed for ease of cleaning. Plate-type or spiral heat exchangers are commonly used. In the laboratory, batch sterilization either *in situ* or in an autoclave handles almost all problems of sterility. At the pilot scale, it is suggested that continuous sterilization and use of heat exchangers be introduced to simulate large-scale operation. While not strictly required, the added equipment (and effort) will result in information on heat transfer coefficients, heat balances, fouling factors, and other operating information which will prove useful, if not invaluable, in later design and plant operation.

V. INOCULUM DEVELOPMENT

Strain improvement begins once a product or expected product is identified. Strain improvement continues through every development and design phase and normally continues, even with reduced emphasis, through the product life. A number of chapters on strain selection and improvement are given in a recent book and in an article by Ritchie.[23,24] Pure culture methods are detailed by Dalby.[25] Maintaining viability and productivity of cultures is not a simple task. No standard procedure can satisfy the myriad of cultures that are in commercial use. A separate sort of optimization scheme is required to determine how best to store and also to rejuvenate the culture. Stability tests of the culture, under varying conditions, should be started early in development. Those storage parameters, which have a marked detrimental impact on viability and productivity, should be identified and precautions taken to prevent culture loss or decay.

In general, insufficient attention has been paid to inoculum development and optimization. For a long period, simply maximizing cell concentration in some "reasonable" cycle was deemed sufficient. Instrumentation and control were more critical and were impressed starting with zero time in the production fermentor. This design philosophy has changed in recent years. The development of inoculum — starting with storage of the culture — is receiving ever greater attention. The introduction of plasmids to bacteria or yeast brings not only desired genetic patterns, but, in many cases, a major magnification of instability. This problem requires study, an understanding of environmental stresses, and protocols for monitoring undesirable changes in the culture.

The culture is evaluated under various storage conditions. It is important that each stage of inoculum development is monitored, not only for later acceptable productivity, but for maintenance of viability/productivity under differing conditions of storage for different times. This point is very important. It is almost a certainty that in a production fermentation process, some delay will occur either in inoculation of a seed stage or in transfer of the final seed stage to production. In both cases, a fully grown culture must be held until some problem in the cycle is rectified. Should the transfer occur and then "hold" the zero hour seed stage with a low cell level? Or should the developed inoculum be held at reduced temperature with a delay in transfer? These questions should be addressed at the scale-up stage or, better, during early process development.

In the lab, transfer of inoculum is essentially instantaneous (tens to hundreds of milliliters are involved). The inoculum biomass may encounter a different environment in a short time interval, but an evaluation of the lab process will occur where undesirable seed transfer conditions are not present. In a production operation, sterile transfer of hundreds or thousands of gallons of inoculum must occur. Even if transfer time is minimized (a desirable goal), there is a finite time interval involved, which may range from 15 to 45 min. First, the transfer header must be cooled; this is accomplished either by a process step (use cold, sterile condensate or sterile air) or by means of the inoculum itself. In the latter case, some reduction in viable count must occur. Second, if an aerobic culture is used, oxygen depletion during transfer is a distinct possibility. The extent of damage should be determined in the laboratory and need for special precautions or design determined early in the design sequence.

VI. FERMENTATION

The development group proceeds with shake flask fermentations and laboratory fermentors to both understand the process and improve it. Synthetic media are often employed to determine the effect of various substrates, cations, and precursors. Even though productivity (rate or yield) may be less than that achieved with natural media constituents, a better appreciation of nutrition and component interactions results. Concurrently, different natural materials are evaluated to supply known and unknown stimulants in an economic manner.

Table 3
FERMENTATION MEASUREMENTS

Physical	Chemical (extracellular)	Biochemical (intracellular)	Biological
Temperature	pH	Cell mass composition	Morphology
Pressure	Redox	NAD/NADH	Size
Agitation speed	Ionic strength	ATP/ADP/AMP	Size distribution
Agitator power	O_2 (gas)	Enzyme levels	Total count
Shear rate, tip speed	CO_2 (gas)	Nucleic acid levels	Viable count
Heat transfer rate	DO_2	Total protein level	Age/age distribution
Heat generation	$D\text{-}CO_2$	Carbohydrate level	Contamination
Gas flow rate	Substrate level	Intermediates level	Degeneration
Liquid feed rate	Phosphorus content	Amino acids	Mutation
Gas humidity	Free N level	Vitamins	Doubling time
Foaming	Total N level		Aggregation
Broth volume	Product level		Genetic instability
Expanded broth volume (density)	Cation level		
Mass	Precursor level		
Viscosity	Nutrient composition		
Level	Intermediates level		
Turnover time	Conductivity		
Osmotic pressure	Inhibitor levels		
Color	Maillard reaction products		
	Amino acids		

Process development and scale-up have been characterized as exhibiting an "iterative approach".[30] As the process continues, changes in microorganism, medium, and the bioreactor are made to be certain that expected responses do, indeed, occur. Containment problems that are relatively simple on a small scale become complex and costly to resolve as scale increases. On-line measurement of critical control parameters and responses is still a developing science. While many physical variables are relatively simple to monitor and control, there are many chemical/biological variables that are more complex. Not all such variables carry equal weight, but monitoring of a variable should not be discounted or eliminated merely because of cost or complexity. There should be a valid technical reason for not using the measurement. A list of fermentation measurements that may be useful is given in Table 3. The list is not indicative of importance (process development will determine need) nor is it meant to be exhaustive. In a modern pilot facility, computer monitoring and control can be introduced early in design. The ability to program variables is important and is responsive to the need to respond to the changing micro- and macroenvironment.

Fermentation systems have been scaled by maintaining geometric similarity. Whether batch or continuous, with or without air-lift designs, vertical or horizontal, there has been a pilot evaluation with some relationship between prototype and production unit. Scale-up compromises are invariably made, but an effort is always made to satisfy biological requirements as much as possible. Maintenance of equivalency on scale-up has many practical difficulties, not the least of which is cost. Even if capital investment could be justified which would offer equivalent mixing pattern, for example, operating and maintenance costs might be prohibitive for the product under study. Obviously, geometric similarity will not assure successful scale-up. A simple example concerns impeller design. Tip speed is directly proportional to diameter and RPM, power input is proportional to RPM to the third power, and diameter to the fifth power. Furthermore, pumping rate is proportional to impeller design (effective area impinging on liquid) and RPM. A balance of needs and compromises based on these needs must be struck in development and scale-up phases.

VII. CONTROL AND INSTRUMENTATION

Newer fermentation plants are highly instrumented and centralized computer control and monitoring are becoming more common. Therefore, development facilities now have a much greater degree of instrumentation complexity than in prior years. Hopefully, these methods of control and on-line analysis lead to more rapid process development and optimization. One can slip into overdesign at this stage. It is important to divide control variables into those which require monitoring (with and without alarm settings), monitoring plus control (plus which type of control), monitoring with recording/controlling, and which algorithms should be followed and calculated. Examples of monitoring only are

1.   Air header temperature and pressure
2.   Level in neutralizing agent feed tank
3.   Temperature in cooling water header
4.   Steam header temperature and/or pressure
5.   Agitator on or off or RPM

Decisions on alarms can be made at this time and frozen at the process design stage. If the neutralizing agent were ammonia, reservoir or header pressure (rather than level) might be monitored. High and low level alarms might be useful to alert operators of an upcoming problem. An indicator bulb for agitator on/off might be sufficient. Of course, any of the "monitored only" variables could be transmitted for inclusion in more complex calculations, such as heat load or OTR. Controlled variables should have some impact on performance whether it be cell growth or product formation. Step change or continuous (programmed) change in key variables should be considered. It is highly unlikely that in a multiphase biological process, a single temperature, pH, or DO_2 level is optimal through the entire cycle.

Computer control is discussed in a subsequent section of this book. Solomons[26] wrote a useful book on fermentation design. Many aspects of good practice were covered and instrumentation was carefully discussed with specific examples given. Fermentation equipment and instrumentation are also covered in a chapter of an excellent book on amino acid biosynthesis.[27]

Kuenzi and Auden[28] start their article on design and control of fermentation processes with an important and simple thought: "The economic viability of a microbial process is determined to a large extent by its basic design and the extent to which its course can be controlled." The review emphasizes secondary metabolites (especially antibiotics) and does point out the fact that design and/or development emphasis must shift with the life cycle of the product. In the earliest design stage of synthesis of a desirable bioactive compound, the key design objective is to maintain a fast track on schedule. The process, at this early stage, will be both simple and probably expensive. One will tolerate expensive raw materials and relatively low yielding purification steps just so long as quality product is available for testing and sale. In the latest or mature phase of the product life cycle, demand may be falling along with profitability. At this point, maximum utilization of existing equipment is the objective, as investment is difficult, if not impossible, and operating compromises are needed in both support (research) and manufacturing. A listing of constraints in fermentation design is also given. Advanced control schemes are discussed as is the need for integrating process development with strain improvement. There are 56 references given which cover many recent findings.

Since scale-up, almost by definition, implies that every aspect of the micro- and macroenvironment cannot be kept constant, it is essential that the *critical* environmental requirements be satisfied. One must determine first just what the critical parameters are, then

Table 4
PROCESS DESIGN REQUIREMENTS

Have no environmental, regulatory, or safety problems

Minimize capital investment
    Have a flowsheet with an economic model
    Product quality known; in-process and product stability known

Design
    Simplify
    Minimize loops, number of steps, inputs
    Minimize novel operations and need for continuing technical input
    Central panel and local control where needed
    Quality of control

Maintain long-term consistency
    Sensitivity to process upsets understood
    Redundant equipment and controls when essential
    Programmed control where possible, data logging
    Corrosion resistance (and regular testing)
    Physical and chemical requirements known
    Raw material availability, consistency, stability

Maintenance of sterility
    Design for ease of inspection and cleaning
    Measurement of critical temperatures
    Free flow and drainage in all pipes and valves
    Weld and gasket specifications and testing
    Sterility of feed system, defoamer, other inputs

Process troubleshooting
    Designed response
    Process reliability
    Back-up controls and alarms (set points)

Downstream processing (see also items above)
    Selection of process steps
    Stability, dosage, quality assurance
    Formulation

just how much they may be varied without moving beyond the unacceptable response point. Interposed on the response curve is the matter of economics, so a response surface results. It is not always possible to select the low cost choice as interactions occur. Furthermore, some element of overdesign may be included. However, it is clear that a quantitative understanding of responses will allow a reduction in uncertainty while maintaining some control of expenditures.

One end result of process development is a listing of process design requirements. While much of the results must be quantitative, certain qualitative relationships should be explained and clarified. A listing of design requirements is given in Table 4.

VIII. DOWNSTREAM PROCESSING

There are many potential process steps in downstream processing (a later section deals more extensively with selected topics) and a preliminary screening should be performed to select unit processes and unit operations that maximize yield and purity within economic constraints. Generally, each additional step will involve some yield loss and an increment in both capital and operating cost. In process selection, need for auxiliaries (such as a solvent recovery system or adsorbent regeneration system) must be included. In process selection and development, the overall picture must be stressed. If fermentation experts are functioning in one area and extraction experts in another, or there is a philosophical rather than a geographic gap, certain broader goals can be neglected. Use of a more costly fermentation substrate (high dextrose equivalent corn syrup rather than molasses or a mixture of amino acids rather than yeast extract) may result in major improvements in isolation. While unit cost in whole broth may not be at a practical minimum, overall production cost might be much lower if less impurity rejection and fewer isolation steps were involved. In the newer fermentations, purification cost is now a multiple of fermentation cost. There is no indication that this trend will reverse itself. Therefore, a search for an economically optimum fermentor broth product will be advantageous. The optimum will be a ''global'' optimum, encompassing total processing including finished product testing.

The overall downstream process may be broken down into two areas. In the first, gross impurity rejection occurs and the product is concentrated. This can be called recovery or crude isolation. In the second area, final product purification involving complete rejection of impurities (to regulatory limits) occurs. The first stages involve larger volumes, larger equipment, and, usually, larger reject streams. A substrate change which increases unit cost of product in broth, but which alters or minimizes a waste stream composition or mass, may be of enormous importance. The ''new'' waste stream may be salable or may be more compatible with conventional treatment technologies.

Crude isolation steps may include one or more of the following: centrifugation, rotary vacuum filtration, evaporation, solvent recovery, ultrafiltration, ion exchange, and crystallization. Certain of these processes may be involved in final purification as well; scale will probably be smaller, but ultrafiltration and crystallization are often used in finishing operations. Process examples for final purification are gel chromatography, affinity chromatography, electrodialysis, use of specific adsorbents, freeze drying. Some of the newer methods are discussed by Lowe.[29] Affinity partition, affinity precipitation, and UF affinity purification are some of the procedures reviewed. Specific examples are given.

An overview of scale-up is given by Van Brunt.[30] The importance of timely and careful selection of process alternatives is made obvious by the fact that investment in a new fermentation-isolation plant might run as high as $50 million. If this is added to the development/regulatory clearance time frame expressed as a cost, the total could easily add up to $100 million. Many millions of dollars could be saved by intelligent choice coupled with speed in process development and design.

IX. RAW MATERIALS AND MEDIA COST

While raw material cost may not be a crucial economic factor in production of a high value biosynthetic product, care and attention to media make-up is essential right from the very start of a research program. Cost, per se, may be of little importance initially, but substrate availability, variability, and stability should be considered at the outset. It is well known that in order to grow, reproduce, and synthesize primary and secondary products of metabolism, microorganisms must have necessary raw materials available. Four major elemental requirements exist: carbon, nitrogen, sulfur, and phosphorus. Hydrogen and oxygen can bring the grouping to six, but the latter two are available whenever an aqueous medium is employed. There is a very wide spectrum of utilizable carbon and energy sources for microorganisms; new and unlikely microbes with novel biochemical ''internals'' are being isolated with some frequency.

Without including all possible substrates, we can list a number of sources for key elements:

Carbon   Carbohydrates, hydrocarbons, alcohols, CO_2, proteins, fats, oils, fatty acids, organic acids

Nitrogen NH_3, nitrate or nitrite ion, amino acids, proteins and their hydrolysates, urea

Sulfur   Sulfate ion, sulfide ion, amino acids (methionine, cysteine, cystine), proteins

Phosphorus   Phosphate ion

A list of the more common media constituents is given below. Many pure compounds can be metabolized by microorganisms, but they would not normally be added purposefully to commercial fermentation media.

Carbohydrate Sources

Glucose (70%) syrup	Cane molasses
Dextrose	Beet molasses
Sucrose	Whey
High DE corn syrup	Corn meal
Enzose or hydrol	Sulfite waste liquor
Starch (various plant sources)	Wood hydrolysis product
Lactose	n-Paraffins

Oils and Alcohols

Soybean oil	Methanol
Corn oil	Ethanol

Nitrogen Sources

Cornsteep liquor	Yeast autolysate
Distillers dried solubles	Ammonium hydroxide (or NH_3)
Cottonseed flour (62% protein)	Yeast extract
Soybean meal (44 or 49% protein)	Fish meal
Brewers' yeast	Peanut meal
Casein (and hydrolysates)	Urea
Meat peptone or milk products	Nitrate, nitrite, N_2 (nitrogen-fixing bacteria)
Thiocyanate, amines, amides, indole	

Waste streams (straw, canning wastes, manure, corn stover) have not been listed, but many reports on their use as substrates exist and specific microbiological effluent treatment schemes act on certain waste products. The major outlet for these agricultural streams has been single cell protein (SCP).

In cell maintenance and reproduction, a relatively large number of complex macromolecules must be synthesized. If a specific secondary metabolite is desired, still other biochemical pathways must be invoked. The major cellular macromolecules involved in cell growth and reproduction include nucleic acids (DNA, RNA), proteins (including enzymes), polysaccharides (including lipopolysaccharides), and various lipids (fats and oils). While biosynthesis is proceeding, cellular energetics must proceed concurrently. ''Respiratory metabolism'' is a general phrase that covers chemical, osmotic, regulatory, electrical, and mechanical work that some cells must perform at one time or another. There are other classes of important elements and compounds which are not metabolized (but may be), but which are equally critical for cell growth, biosynthesis, and reproduction. These are cations, anions, vitamins, and cofactors. As a generalization, one may conclude that a fermentation medium should contain a carbon source (crude or purified sugar or hydrolyzed starch), a nitrogen source (a yeast product or an agricultural product), and buffer salts. Specific cations, vitamins, and cofactors may be required. A review describing many potential substrates is available along with world production volumes and market prices for selected raw materials.[31] Hydrocarbon and alcohol fermentations are also discussed and potentials for SCP and ethanol are reviewed.

Creating a "soup" of ingredient constituents to cover every contingency for the producing microbe may seem an advisable start; this is illusory. Adding too many ingredients may (1) add repressors or (2) allow component interactions. Repressors are compounds which inhibit certain enzyme reactions; inhibition may occur at a number of sites. Catabolite repression was recognized rather early in studies of antibiotic synthesis; freely available carbohydrate was utilized vigorously and antibiotic synthesis was slowed or eliminated. Interactions between certain cations (Fe, Mg), ammonium ions, and phosphate are known. Maillard reactions between amino acids or peptides and carbohydrates have resulted in formation of inhibitory compounds. Nitrogen catabolite repression has been recognized. Excess is not always the best route in media formulation.

One logical procedure to use in formulating a medium is to review in-house and published information. The organism of choice or a related one may have been successfully grown on a more or less defined medium. In a strict sense, a "defined" medium is one in which every compound added is quantitatively and qualitatively known. The detail in a defined medium would give names (and purities) of known compounds and their respective composition (concentration). More loosely, a medium may contain natural products with components falling into a range of concentration. It is no longer a defined medium. A parallel approach is to try a material balance. Cellular stoichiometry is sometimes known; very often, it can be inferred from data on related species. Nutrients are then blended in sufficient quantity and in those proportions to give a desired and known level of biomass. Usually, some margin of error is allowed and a slight excess is added. If noncellular products are desired (and intracellular storage products are here included), the component balance calculation must include substrate-to-biomass and substrate-to-product conversion. The efficiency of incorporation (the "yield") of substrate to cell or to product must be known with some degree of certainty. In many cases, a defined medium supports growth or product formation. Complex mixtures need not be added, but it often happens that addition of a natural product stimulates growth or production by a factor beyond that which can be explained by the composition of matter alone. The natural product must be deemed nonessential since it is not required for growth; still, these crude and relatively inexpensive ingredients supply both known and unknown factors to the cell. A library of large volume ingredients exists for use in fermentation media. Many variations will fulfill growth and production requirements; not all will be equally efficient, but a general plateau of efficiency should exist. Only rarely will media manipulation cause an abrupt and stepwise increase in yield provided no repressors or inhibitors were present. Still, the well-known story of cornsteep liquor stimulation to penicillin fermentation acts as a beacon to investigators.

The elemental composition of microorganisms depends upon many factors: type of microbe, substrate, batch or continuous culture (also dilution rate), and operating parameters. Some typical ranges may be projected with the understanding that not every cell or microbial system will fall into each range. If a cell is presumed to be 80% water, the other 20% consists of 11.5% protein (of this value 10 of the 11.5 is nonribosomal), 5% RNA, 1% each of DNA, polysaccharide, and lipids, and 0.5% of other, small molecules. The range of elemental analyses is (percent by weight):

	%
Carbon	45—53
Hydrogen	5—8
Nitrogen	9—14
Oxygen	26—35
Ash	8—9

Table 5
**COMPOSITION OF MEDIA USED FOR SELECTION OF
GLUTAMIC ACID-PRODUCING BACTERIA (VALUES OTHER
THAN pH IN %)**

	1	2	3	4	5	6	7
Glucose	5	5	5	1	5	2	5
$(NH_4)_2SO_4$	—	2	—	—	—	—	—
$(NH_4)_2HPO_4$	—	—	—	—	0.3	—	—
Urea	0.8	—	0.8	0.5	0.5	—	0.5
Meat extract	0.2	0.2	0.2	—	—	—	—
Peptone	0.2	0.2		0.2	—	—	—
Yeast extract	—	—	—	0.5	0.5	—	—
Cornsteep liquor	—	—	—	—	—	0.5	0.5
Salts[a]				Add in all cases			
$CaCO_3$	—	3	—	—	—	—	—
pH	7.2	7.2	7.2	7.2	6.0	6.0	Not given

[a] Salts include K_2HPO_4 (0.05—0.1%) and $MgSO_4 \cdot 7H_2O$ (0.02—0.05%).

From Kinoshita, K., in *The Microbial Production of Amino Acids*, Yamada, K., et al., Eds., Halsted Press, New York, 1972, 265. With permission.

A number of "typical" fermentation media for various processes are discussed below to show similarities and differences in formulation. References noted may be consulted for many more media compositions. Another excellent source of media composition is the patent literature. One key point in development is that as strain improvement proceeds, nutrient requirements may change. Repression by glucose and by phosphate is well documented for many antibiotics, and levels of nutrients that are suitable for an earlier culture may be totally unsuitable for a mutant or recombined culture. Catabolite repression (carbon or nitrogen or phosphorus) can be countered by using a slowly metabolized substrate or by controlled feeding. In some cases, complexes may be formed or added and inhibition is prevented. A search for inducers or mutants which are resistant to inhibition continues. As metabolic pathways are better understood and controlled feed or additions are made at the pilot scale, major improvements in raising product titer can be achieved. The point here is that media development is an ongoing process, continuing as long as the process is commercialized.

In the mid-1950s, Japanese workers performed massive screening of microorganisms to isolate those that produced glutamic acid. The variations in media are instructive; data are given in Table 5 (taken from Reference 27, p. 265).

It must be remembered that biotin is an essential growth factor for GA-producing bacteria. The optimal concentration is strain dependent. In the same volume, various media are described in screening studies for production of other amino acids. Valine fermentation was susceptible to effects of heavy metals; therefore, levels of many metal salts were varied to determine not only direct effects, but interactions. This example of multimedia evaluation illustrates another important point. In early stages of development, it is unlikely that the medium is optimized. A semiempirical variation in media constituents as well as other parameters can result in marked improvement even at the shake flask scale.[24] In cases where a precursor is required, optimum levels must be determined. Normally, precursor cost is relatively high on a unit basis, and cost-benefit relations must be determined. One error sometimes made in evaluating precursor or stimulant use is that of viewing fermentation economics only. The unit cost of product in the harvested whole broth may, in fact, rise with use of some costly stimulant. As such, its use may be questioned even if product concentration rises. What must be included is impact on *total* process economics. Down-

Table 6
PRODUCTION MEDIA FOR MICROBIAL
VITAMIN PRODUCTION

Vitamin B$_2$
Organism:
Ashbya gossypii

	% (w/v)
Cornsteep liquor solids	2.25
Wilson's peptone W-809	3.5
Soybean oil (Glycine supplemen-tation at 1—3 g/ℓ was shown to be stimulatory)	4.5

Vitamin B$_{12}$[a]
Organism:
Pseudomonas denitrificans

	g/ℓ
Beet molasses	100
Yeast	2
$(NH_4)_2HPO_4$	5
$MgSO_4 \cdot 7H_2O$	3
$MnSO_4 \cdot H_2O$	200 mg
$Co(NO_3)_2 \cdot 6H_2O$	188 mg
5,6 DMB[b]	25 mg
$ZnSO_4 \cdot H_2O$	20
$Na_2MoO_4 \cdot 2H_2O$	5

[a]   From French Patent 2,038,828, December 28, 1970.
[b]   5,6 DMB is 5,6-dimethylbenzimidazole.

stream processing costs may change very little, if at all, while purifying the "higher" cost broth. Throughput rises, therefore, and overall product cost might decline significantly due to a reduction in unit overhead absorption.

Vitamins B$_2$ (riboflavin) and B$_{12}$ (cyanocobalamin) are produced by fermentation. Media preparation is related to the producing microorganisms' needs as well as the structure of the desired product. See Table 6. Operational details can be found in separate articles in a single book.[32]

Examples of antibiotic production media show some general tendencies:

Erythromycin[33]	Glucose 40 g/ℓ, corn extract 10 g/ℓ, $(NH_4)_2 SO_4$ 6 g/ℓ, NaCl 2.5 g/ℓ, $CaCO_3$ 10 g/ℓ, sperm oil 6 g/ℓ
Bacitracin[34]	Cottonseed and soybean meal, $CaCO_3$, dextrin; or soybean grits and sucrose
Chlortetracycline[34]	Sucrose 1%, CSL 1%, $(NH_4)_2HPO_4$ 0.2%, KH_2PO_4 0.2%, $CaCO_3$ 0.1%, $Mg SO_4 \cdot 7H_2O$ 0.025%, $Zn SO_4 \cdot 7H_2O$ 0.005%, $CuSO_4 \cdot 5H_2O$ and $Mn Cl_2 \cdot 4H_2O$ each at 0.00033%; or sucrose, peanut meal, CSL, molasses, $(NH_4)_2SO_4$, $CaCO_3$, NaCl

A. Sugar Sources

An excellent and concise summary of world production and price of sucrose and molasses can be found in an article by Brown.[35] World production of sucrose is *circa* 100 million t/year. Molasses is defined as the by-product of sugar refining from which no additional sucrose may be crystallized by conventional means. Total volume of molasses is about 30 to 35% of sugar production. While molasses quantity is related to sugar volume, there is no direct relation in price. About 80% of world molasses supply is used locally; the remaining material is traded internationally. Molasses production in the U.S. is roughly 1.5 to 2 million t/year with a further 1 million t/year imported. There is a weekly price quote (dependent upon port location) prepared by the Department of Agriculture; the quote (see below) reflects spot prices in a rather special form and is used as a world market price indicator. Molasses prices (cane, ex-tank, New Orleans) ranged from $40 to 50/short ton from 1975 to 1978, rose to about $82/short ton in 1979, and close to $100 in 1980. There has been a decline since then. While fermentable sugar availability ranges from 45 to 60%, a value of 50% can be used for calculation purposes. For molasses delivered at a plant site for $60/short ton, sugar cost is $0.06/lb. The U.S. sucrose market is protected and supported price is in excess of $0.20/lb; certain limited and stringent exemptions exist and there must be a strong and valid reason for using sucrose in place of molasses as a carbohydrate source. The price of both sucrose and molasses is considerably lower in less developed countries compared to the U.S., Europe, or Japan, but neither the capital nor infrastructure exists to take full advantage of this raw material cost reduction.

As noted above, about one third the mass output of sucrose is equal to the weight of molasses produced. In fact, the world molasses production estimate for the 1983/84 crop year is 33.3 million metric tons.[36] Brazil, India, and the U.S.S.R. produce about one third of world molasses output. Production of molasses world-wide has shown a plateau for the past 3 years (1981/82, 1982/83, and 1983/84). Where full year data are available, here is the recent pattern of molasses usage in the U.S. (units are millions of gallons of molasses):

	1980	1981
Distilled spirits	8.2	4.9
Pharmaceuticals and other edibles	18.8	19.1
Yeast, citric acid, and vinegar	77.5	75.2
Animal feeds	364.3	425.6
Total	468.8	524.8
Imports	140.3	195.9

For the 1982/83 year, molasses imports rose to an estimated 310 million gallons. The average wholesale price of blackstrap molasses (BSM) at New Orleans is shown in the following table:

Year	$/ton
1979	83
1980	96.40
1981	76.30
1982	47.60
1983	56.45
1984	61.55

The price varies dependent upon port of entry and local marketing conditions. For example, FOB price for tank car or truck, California ports is approximately $4 to $5/ton higher than the New Orleans posted price. These data are updated and reviewed by the U.S.D.A. in Denver in a weekly report and annual summary of the *Molasses Market News*.

In stark contrast, raw sugar on the New York spot market (CIF, duty paid) averaged

$0.2204/lb for 1983; the price was essentially the same (to 0.1 mil) in April of 1984. As noted, the U.S. domestic sugar market is well protected. The average spot raw sugar International Sugar Agreement world price was $0.0846/lb in 1983 and even this value declined in the first half of 1984.

Corn is abundantly available in the U.S. and a mix of sugar sources is derived from corn starch. A review of technology and economics is available.[37] Many streams derived from corn processing are usable in different fermentations; there is no simple predictor as to which process requires a higher purity stream. A flow diagram and material balance of the corn wet-milling process are available.[38] For every 100 lb (d.b.) of shelled corn, 67 lb of starch and sweeteners are produced plus 7.1 lb of cornsteep liquor. One item of commerce is dextrose monohydrate, which is a white, crystalline powder containing 91.5% anhydrous dextrose. A liquid dextrose stream (71% solids) is available for tank truck or railcar delivery. The carbohydrate present is essentially all glucose. Many syrups of varying degrees of hydrolysis are commercially available. A high DE (dextrose equivalent) syrup means that essentially all the starch has been hydrolyzed to monomer. In the last decade, high fructose corn syrups (HFCS) in massive volume have been made available for commercial use. Enzymatic isomerization results in syrups containing 43 or 55% (of total solids) fructose. The price of HFCS has historically been 10 to 20% below that of sucrose; inroads, especially into soft drinks, have been remarkable. Finally, a material that can be dubbed "corn molasses" (or hydrol) is available. Just as with "regular molasses", no further carbohydrate crystallization can occur using conventional means. Hydrol has limited use as a fermentation substrate; however, its price on a solids basis is attractive. Price history for some corn sugars is given below ($/kg):

Year	Dextrose	HFCS	Corn syrup
1978	0.33	0.19	0.18
1979	0.35	0.21	0.21
1980	0.59	0.38	0.32
		42% fructose	SG = 1.42
		71% wt solids	80.3% wt solids

The review article compared gross composition of various molasses:[37]

	New Orleans cane	Wisconsin beet	Corn molasses
Solids, %	80.8	78.6	74.9
Total sugar, invert, wt %	59.5	52.7	50.3
Crude protein, %	3	11.4	0.4
Ash, %	7.2	9.3	8.9
pH range	5.5—6.5	7.5—8.6	

Of some importance is the fact that cane molasses not only has a lower ash level than beet, but the former contains higher vitamin levels than the latter.

Dried (or liquid) whey offers relatively low cost and supplies not only carbohydrate (lactose), but also protein and a number of vitamins. One typical analysis of dried whey shows:

Table 7
AVERAGE VALUES FOR CONSTITUENTS OF
OILSEEDS AND MEALS

Values are Percents, Unless Noted

	Soybean		Rapeseed		Sunflower	
	Seed	Meal	Seed	Meal	Seed	Meal
Protein	38	45	24	43	16	50
Ether ext. fat	20	0.8	40	2.3	45	3.1
N-free extract		33.8		36.8		26.7
Fiber		6		10.7		11.6
Ash		5.8		7		8.3
Ca		0.3		0.8		0.3
P		0.6		1.2		1.2
Metabolizable energy (Mcal/kg)		3.6		3.1		2.9

Note: The chemical score for sunflower protein is 56% due to lysine deficiency. Rapeseed protein chemical score is 90%, while the equivalent value for soy protein is 91.5% (methionine and cystine limitation).

			mg/lb
Carbohydrate	68%	Riboflavin	9
Ash	9.6%	Panthothenic acid	22
Solids	95%	Niacin	5
Protein	12%	Pyridoxine	1.3
		Choline	1.1

Dried whey price fluctuates with the season and can easily change by a factor of 2 over 1 year. Prices in recent years have been as low as $0.10 and as high as $0.25/lb FOB. The source of whey is cheese making so the greatest concentration of producing locations is in Wisconsin, New York, and California. Unconcentrated (*circa* 6 to 7% solids) whey may be available at very low cost or at zero cost at a producing location. Cost of haulage is the material cost. The obvious requirement is for a lactose-utilizing (or hydrolyzing) microbe; otherwise, initial hydrolysis to glucose and galactose is required. Such pretreatment adds to material cost.

B. Agricultural Products as Nitrogen Sources

The predominant oil seed in world production is the soybean. Lewin gives simple flow charts for soybean processing (to oil and meal) and procedures for refining the oil.[39] About half the world oil seed production is soy and about 60% of world soybean production occurs in the U.S. A detailed history of soybean crops is given for 1971/83 for the U.S., Brazil, and Argentina. U.S. soybean supplies, acreage, and yield are also given for the 1970/83 crop years. In a normal year, the price of soy meal is determined by demand as an animal feed. Since an animal feed can (and is) produced to a final specification in terms of energy, protein content, and cost from an almost infinite number of raw materials, soy meal must compete with other natural products (animal, vegetable, microbial in origin). The selling price is a continuously changing function of availability and international demand.

Soy supplies 80% of the world demand for protein cake and 25% of the vegetable oil trade.[40] The overall utilization of defatted, high protein seed meals is overwhelmingly as

Table 8
NITROGEN SOURCES FOR FERMENTATION MEDIA

	Percent N by weight	As is ($/kg)[a]	Approx ($/kg N)
Ammonia	82.4		
Barley	1.5—2		
Beet molasses	1.5—2		
Brewers' yeast	8	0.45	5.62
Cane molasses	1.5—2		
Cornsteep liquor	4.5	0.20	4.44
Cottonseed meal	5.2		
Distillers dried solubles	4	0.15	3.75
Oat flour	1.5—2		
Pharmamedia	8	0.55	6.88
Rye flour	1.5—2		
Soybean meal	8	0.22	2.75
Whey (dried)	4.5	0.42	9.33

[a]   Includes $0.02—0.04/kg shipping cost.

animal feed (some 98%). The critical constituents of three oilseed meals are given in Table 7.

Another reference gives details on production of soybean meal and cottonseed meal, as well as many other agricultural products.[41] Detailed amino acid compositions are also given. Since soybean meal is traded as a commodity, an annual report gives a voluminous amount of price and production data for the U.S.[36] The U.S. has accounted for 38 to 43% of world soybean meal (SBM) production, but this value dropped to 36% (or 20.9 million t) in the 1983/84 season. Average price of 44% protein SBM (Chicago, bagged) in $/ton shows this history:

Year	Av. price
1978	182.59
1979	202.76
1980	209.33
1981	215.96
1982	192.07
1983	211.50
1984	166.84 (Decatur, bulk)

Futures prices for 1985 have dropped by 30% and more giving a decade-low figure. This reflects reduced feed use demand, lower purchases by Eastern bloc countries, and good supply world-wide. Other feed products whose prices are tied to SBM price (such as lysine) show a similar price drop.

There are many other potential sources of organic nitrogen; these sources are compared to ammonia in Table 8.

Selected characteristics of five organic nitrogen sources which have been used in commercial fermentation are given in Table 9. The values for fishmeal are given as ranges; concentrations are dependent upon source of the fish. Data are abstracted from Considine.[41]

Table 10 lists selected typical analyses of various fermentation media constituents. There are some discrepancies in fiber content, but this is not a critical item for the intended use

Table 9
CHARACTERISTICS OF SELECTED NATURAL INGREDIENTS (VALUES IN PERCENT)

	Dry matter	Crude protein	Crude fat	Crude fiber
Corn distillers dried solubles	93.8	27	9	13
Cottonseed meal, solvent, 40%	91.4	41	4	13
Fishmeal	91—93	62—72	10	1
Soybean meal, 44%	89.6	44	0.5	7
Meat and bone meal, 50%	92.6	50	8.5	3

Conc./kg[a]

	Vitamin A	Vitamin E	Vitamin B12	Vitamin B2	Niacin	Pantothenic acid	Choline	Folic acid	Thiamin B1	Pyridoxine
CDDS	0.8	82	29	17	115	21	4818	1.1	6.8	11
CM	NA	13	NA	5	40	31	2867	2.2	6.6	6.6
FM	NA	4—22	185—212	6.6—9	55—88	8.8—11.5	3087—4037	0.2	0.7	2.6—3.5
SBM	NA	4.4	2.2	3.1	22	13.2	2751	0.7	2.2	7.1
MBM	NA	2.2	44	3.7	42	3.5	2183	0.4	1.1	1.5

[a] Vitamin A in 1000 IU; vitamin E in IU; vitamin B_{12} in μg; others in mg.

Table 10
TYPICAL ANALYSES OF INDUSTRIAL MEDIA CONSTITUENTS[33]

	Corn meal	Cornsteep liquor	Cottonseed meal	Distillers dried solubles	Soybean meal	Brewers' yeast
	%					
Protein	8.9	24	41	26	51	43
Carbohydrate	68.9	5.8	28	45		40
Fat	3.9	1	1.5	9	1	1.5
Fiber	2	1	13	4	3	1.5
Ash	1.3	8.8	6.5	8	5.7	7
Solids	85	50	90	92	92	95
	mg/lb					
Riboflavin	0.6	2.5	2	7	1.4	16
Thiamin	2	0.4	6.5	2.5	1.1	34
Niacin	9.7	38		50	9.5	227
Biotin	0.03	0.4		1.3		0.5
	%					
ℓ-arg	0.4	0.4	3.3	1.0	3.2	2.2
ℓ-cys	0.2	0.5	1.0	0.6	0.6	0.6
ℓ-gly	0.4	1.1	2.4	1.1	2.4	3.4
ℓ-hist	0.2	0.3	0.9	0.7	1.1	1.3
ℓ-isol	0.5	0.9	1.5	1.6	2.5	2.7
ℓ-leu	1.2	0.1	2.2	2.1	3.4	3.3
ℓ-lys	0.3	0.2	1.6	0.9	2.9	3.4
ℓ-meth	0.3	0.5	0.5	0.6	0.6	1.0

and methodology for fiber content varies considerably. Selected raw material pricing is given in Table 11 and a comparison is made to data presented a few years ago.

In cost determinations relating to raw materials, some important generalizations should be considered.

1. Material cost increases as purity rises. There is usually enough crude natural product in the medium to allow use of technical, rather than reagent or USP, grade salts.
2. By-products of other processes will usually result in the lowest cost of raw material for fermentation.
3. Contract prices will invariably be below published or spot prices. Terms will depend upon time period, quantity, quality, penalties, and may also be dependent on selected commodity price fluctuations.
4. By-products made in course of production should be taken as a credit, but only at a real market price, and after subtracting necessary processing costs and freight, if applicable.
5. Many materials (such as ion exchange beads, ultrafiltration membranes, adsorbents) must be replaced at a known or at least predictable frequency. The periodic make-up cost for the agent is an operating cost and may be included as a raw material. If not included here, it must be included elsewhere as a variable cost.
6. Freight costs must be included whenever applicable.
7. For animal or plant tissue culture, many high purity and costly substrates may be required. Special pretesting or pretreatments may be needed. Raw material costs may be one or several orders of magnitude higher than those incurred in "conventional" fermentation.

Table 11
RAW MATERIAL PRICE COMPARISON

	Market estimates ($/kg)	From Swartz[135]
Dextrose, bagged	0.48	—
Glucose, 70% w/v	0.31	0.154
Enzose, 72% w/v	0.18	0.099
Sucrose	0.51	0.44
Beet molasses	0.112	0.081
Starch, pearl	—	0.165
Lactose	0.34 (tech.)	0.44
Corn syrup, 95 DE	0.36	—
Corn meal	0.26	—
Soybean oil, refined	0.705	0.606
Corn oil	0.715	—
Ethanol	0.60	0.366
Soybean flour	0.21 (meal, bag)	0.265
Cottonseed flour, 62% protein	0.55	0.43
Cornsteep liquor	0.195	0.11
$(NH_4)_2 SO_4$, by-product	0.13	0.254
Distillers dried solubles	0.15	—
Brewers' yeast	0.45	—
	Mid-1985 delivered midwest	1979 Delivered to Indianapolis, Ind.

A typical calculation is given to show how raw material cost can be derived. Citric acid fermentation is the example and a hypothetical medium is given:

	g/ℓ	Unit cost ($/kg)	Usage (kg/kg citric)	Product cost ($/kg)
Molasses, beet	200	0.0814	2.50	0.2035
Potassium ferrocyanide	0.6	2.64	0.0075	0.0198
KH_2PO_4	0.2	2.156	0.0025	0.0054
$MgSO_4 \cdot 7H_2O$	0.25	0.308	0.0031	0.0010
$ZnSO_4 \cdot 7H_2O$	0.005	2.97	6×10^{-5}	0.0002
NH_4NO_3	2.0	0.165	0.025	0.0041
Methanol	20	0.186	0.25	0.0465
				0.2805

In the case shown, the harvested broth was assumed to contain 80 g/ℓ citric acid (80% yield based upon 100 g/ℓ sugar present initially). If broth contains 60 million lb citric acid on an annual basis and fermentor yields were 90 instead of 80%, a savings in raw material cost of $1.87 million would occur. Reduction or elimination of methanol from the medium would also have a positive financial impact of similar magnitude. The product cost above is that in harvested broth; it should be divided by the extraction efficiency to give a fermentation raw material cost on a unit product basis (packaged or loaded). To that figure, add cost of extraction/purification chemicals (on a consistent basis) to give a total materials cost. Total materials cost is normally given per unit of finished product; this may be a weight, volume, or dosage. Packaging costs are, therefore, included unless otherwise noted.

The cost of sugar is rather critical in biosynthesis of a commodity item.[42] The mid-1985 EEC price for sugar is about 3.5 times the world market price. (In the U.S., this ratio is variously 2.5 to 3 times the world price.) Whereas in antibiotic production (unnamed product) feedstock cost is 2.5% of total production cost at world sugar prices, the feedstock cost rises to about 8.5% of total cost at EEC pricing. This is of some importance, but probably tolerable for an antibiotic. For citric acid, the cost of sugar would be about 36% of "product worth" (given as approximately $0.58/lb) at world sugar prices. However, at EEC sugar pricing, cost of sugar would be *1.28 times* the product worth. The situation is comparable for polyhydroxybutyrate polymer. Whereas plant location is not always dependent upon raw material cost, availability and cost are obviously very important determinants.

In summing up this section, medium selection and preparation involve these selections:

Suspending medium   Usually water, but hydrocarbons and alcohols have been or are in use

Carbon source   Initial content and feeding levels

Need for pretreatment, digestion, hydrolysis: normally not required as cost factor, at this stage may be unacceptable

Nutrients/activators   Nitrogen (inorganic plus organic), phosphorus, sulfur, cations, vitamins, cofactors

Precursors/inducers   If required in biosynthesis (as in certain antibiotics or vitamin B_{12})

Defoamer   Minimize toxicity or inhibitory effects

pH control   Buffers, internal or added; caustic soda, soda ash, lime, bicarbonate, ammonia, or urea; potassium hydroxide

Protective agents   Complexing agents, antibiotics (protective additives)
(only if essential)

The selection criteria involve the microbe and its metabolism, of course, but other factors are cost, availability, ease of handling, storage requirements, downstream processing, and waste disposal considerations.

Finally, there is an exhaustive review article on raw materials for biotechnology.[43] Dimmling and Nesemann discuss crop and cereal production back to 1970. The last decade of crude oil and n-paraffins prices is shown. Grains, wood, and bagasse are detailed, as is pricing for sugar and molasses. Detailed compositions are given. Many complex natural products are listed with very complete analyses (including minerals, vitamins, amino acids). Typical media compositions are given and 172 references are listed. The article should be consulted by anyone involved in fermentation media selection and optimization.

X. PROCESS DESIGN

Once process development and scale-up data have been compiled, a process design package must be prepared. The goal of process design is a documented outline, set of specifications, and procedures which present system details that will permit selection of equipment, controls, and services to effectively carry out the developed technology. Included are mass and component balances, energy balances, flow rates, specifications, installation drawings, and operating parameters.

Designs can be preliminary (usually done early in project cycle even before definitive data are developed), detailed (done after a reasonable amount of hard data are developed), or firm (the definitive or detailed package). Cost for each type rises proportionally with effort and detail. Potential for faults in process design and error in investment cost estimate fall as the package becomes more detailed. The design basis is given below in outline form. Details of seed development and fermentation have been described earlier; the format here is basically that of a chemical reactor (time, temperature, process parameters, inputs, and outflows). Certain additional details for other design items are listed, since these details may receive cursory review. A number of process design questions must be raised at this stage of the project and raised again during the engineering design phase. Hopefully, the answers will not have changed in the interim; however, they become more firm with the passage of time and expenditure of design dollars. There is a natural flow and overlap in the design cycle. ''Process design'' is meant to encompass earlier phases, while ''engineering design'' refers to the final phases.

Preparation of Estimate (Basis for Design)

Process
  Raw materials
    Types and form
    Receiving and storage
    Transfer and inventory control (mass balance)
  Inoculum preparation
    Sterility requirements, isolation, containment
    Space, equipment, access
  Seed development
  Medium preparation
    Batching
    Continuous feed
    Defoamer
    Sterilization
  Fermentation
  Harvest
    Method and volume (mass) measurement
    Stability
    Special treatments
  Downstream processing
    Regeneration, recovery, replacement (resins, adsorbents, etc.)
    Containment
  Waste disposal
    By-product streams
    Contaminated batch (or other failure)
    Impact on air, ground, water
    Liquid or solid disposal and sites
  Blending, packaging, dosage forms
    Potential for cross-contamination
    Testing
  Product storage
    Temperature, humidity
    Stability
  Operating factors, rates, yields (stepwise and overall)

Nonprocess
  Process documentation
    Regulatory requirements
    Computer storage/hard copy
  Utilities
    Air
    Water

        Steam
        Electricity
        Storm sewer
        Sanitary systems
        Plant drains
        Fire protection
        Standby systems
    Quality control laboratory
        Regulatory requirements (paper/computer storage)
        Critical control points
    Personnel facilities
        Uniform requirements (isolation areas)
    Administration facilities
        Access
        Support laboratories
    Maintenance and stores
    Warehousing, inventory control, rework, transportation

For all these criteria, there is one fundamental and overriding requirement. Safety requirements must be satisfied. Any hazardous materials or steps must be fully documented and essential data transmitted with process information. The quality of a process design package not only involves rate, yield, cycle times, and so on, it must involve safe operation for all involved in and near the producing plant. In a developed process design package, safety is truly first.

If multiple products are to be made, there are new approaches to optimization. Alternate arrangements may be simulated and most efficient use of storage, intermediate holding, and process equipment can be projected.[44]

The detailed design estimate should be prepared only after all process and production factors are fixed or established within a very narrow band, otherwise the result will not be worthy of the effort. Examples of factors which should be firm are

1. The manufacturing process including mass, component, and energy balances
2. Temperature, pressure, pH, control ranges, and hazards analysis
3. Raw material, in-process and finished goods, specifications, and testing procedures
4. Time cycles, turnaround, on-stream factors, yield, rate, annual output
5. Utilities requirements and energy availability
6. Materials of construction (corrosion data for all process streams including waste streams)
7. Regulatory requirements such as GMP, environmental issues
8. Retrofit, add-on or new plant site (see Reference 45)
9. Transportation (loading/unloading) requirements
10. Design for potential catastrophe (fire, flood, earthquake, severe climate)
11. Building and structural design (include codes), health, and safety
12. Storage, including in-process and product

One design criterion which is often neglected could easily be placed first, i.e., "Design for future expansion". For commodity chemical products, location of a plant near markets is important. The factor is less important for a highly active pharmaceutical.

Before moving to detailed charts and drawings, it may be useful to take a broader view of a fermentation process and analyze it as a specific type of production operation. Fermentation schemes have a number of distinct characteristics and the process design philosophy should be established to be responsive to these needs and characteristics. Scheduling,

Table 12
CHARACTERISTICS OF INTERMITTENT FERMENTATION PRODUCTION SYSTEMS

Similar operations are grouped	Individual work center schedules must be developed and co-ordinated; multiple batch fermentors allow for many products harvested in sequence; purification areas are used at different times
Products are diverse, often custom designed	Raw materials supply, preproduction planning needed; scheduling may be weekly, monthly, or quarterly with demand variation; process change is common
Processing steps may be uncoupled and changes in flow may occur	Relatively complex control system required to monitor flows, purities, stability; monitoring of critical control points essential
In-process inventories may build up	Flexibility in scheduling required due to process upsets, contamination, mechanical downtime, change in demand; in-process streams may be very unstable
Work loads unbalanced between processing steps	High and low labor/utilities demand exists; scheduling must adapt to fully load facility
Mechanical or electrical breakdowns do not immediately affect subsequent steps	Sequential operations plus surge capacity allow for downstream processing to continue for a time; degradation of held material is likely, so some redundancy is built in; a fermentor loss is costly, not only at that stage but in idled equipment in extraction plant
Products typically of produce-to-order type	Relatively long lead times needed both for processing needs and quality checks; physiological testing may be required plus special dosing needs; scheduling is complex to assure minimum cost production

movement of materials, spacing of equipment and buildings, testing requirements, and control needs all contribute to setting a process design *tone*. The processing steps in a fermentation-purification train are often unique. While general engineering principles can and should be applied, special requirements should be satisfied to maximize efficiency, minimize loss or downtime, and therefore keep costs to a minimum. The characteristics of an intermittent fermentation production system are given in Table 12. These would require some modification for a continuous system or for a tissue culture process, but the overall concept is clear. The quantitative results should satisfy needs for mass and heat transfer, but should simultaneously satisfy the qualitative needs of the process. The design will be set to minimize disruption or maximize rate and yield, after a thorough review and incorporation of these process characteristics.

Each design element will be reviewed with a view toward points sometimes overlooked or neglected.

A. Raw Materials

Major raw materials have been reviewed earlier. Form, packaging, minimum purity, and storage conditions will be known. Inventory levels and two to four suppliers per item will be identified. Domestic supplies are preferred. For certain high volume natural products, contacts on contractual arrangements will have been initiated. Methods of verifying mass or volume levels at a fixed time (inventory needs) will be under consideration or selected. Monitoring of transfers and means of transfer (raw material movement into and within plant) will be planned. While large volume items receive most of the attention, all raw materials — even those lumped as operating supplies — must be considered. If some treatment or cleaning chemical is used at a rate of 10,000 gal/month, it is necessary to consider an optimum size storage tank. What are price differentials for different volume deliveries? What are desired inventory levels? What are storage (tank) costs and material stability? The same review is needed for solid materials and supplies.

Many materials are toxic or hazardous (acids, bases, fuel oil). Written documentation is a regulatory requirement. Major changes are being made in laws concerning underground tankage; new monitoring rules are being formulated for existing tanks and new construction regulations are being implemented for newly installed tanks. For aboveground tanks (including those for chlorine and ammonia) special requirements are being introduced to prevent catastrophic discharges. Special diking or other containment procedures must be considered in the design phase.

Alarm and monitoring systems should be considered at the early design phase; add-ons will cost more and may interrupt operations. The storage and transfer of raw materials must include design considerations involving:

1. Results of catastrophic rupture or loss (groundwater, nearby communities, waterways as streams, river, ocean, soil contamination, sewers)
2. Containment, dikes, channels
3. Effect of leaks on underground piping and electrical lines
4. Fire or explosion hazard, remote indication
5. Contamination by rodents, birds, insects, foreign objects
6. Location of overhead pipes and lines and interferences to delivery and in-plant transfers
7. Interconnections (for vessel cleaning, repair, mixing)
8. Neutralization potential (examples are caustic availability to neutralize an acid leak, or water spray to absorb escaping chlorine)
9. Proper signs to indicate hazards (multiligual)
10. Availability to railcars, barges, trucks (one or combination)
11. Seasonal availability, especially of natural products
12. Conditions of storage (GMP or regulatory review)
13. Protection against riot or sabotage

One other item is often overlooked. In biosynthesis of a regulated product such as a drug, substitution of a less costly raw material or solvent often requires approval of a regulatory body. If possible, flexibility should be built in *early* in the approval process.

B. Inoculum (Seed) Preparation

At this step of design, many criteria are known. Design effort is centered upon laboratory needs, equipment, degree of isolation, and movement into and out of the inoculum laboratory. It is generally considered wise to have a separate and distinct inoculum storage, development, and testing area or building. Ingress should be limited, with more severe restrictions as one moves to "cleaner" areas (meaning those rooms or areas having lower and lower volumetric microbial counts).

Control of contamination is probably one of the "key" design criteria. There is a small number of chemical processes that have such stringent requirements for prevention of entry of foreign organisms. As one moves from the inoculum lab to the plant, one loses the enclosure protection (controlled entry, laminar flow hoods and rooms, controlled and filtered airflow, maintenance of a positive pressure) and enters a different environment. Inviolability of the process vessels, inputs, and interconnecting piping must be designed. Valves, piping, vessels, internals, and sensing instruments must be designed for ease of cleaning, free drainage, steam or chemical sterilization, no (minimal) corrosion, and ease of maintenance. Vessel-to-vessel transfers must be accomplished simply with assurance of short pipeline residence times. The seed cell mass should not undergo undue temperature rise, substrate or oxygen depletion, or pH change during transfer.

Seed vessels may be any size, but inoculum ratios are usually 5 to 15% so the seed train is designed accordingly. There has been a tendency to eliminate the smallest sparged fermentor and move from flask to "large seed". The additional time in the "large seed" to reach optimum cell mass is more than compensated by elimination of one stage and one transfer. The seed unit must be provided with capability to chill the broth; it may sometimes be necessary to hold fully developed inoculum because of a scheduling problem. Seed vessels are normally manifolded so a single line is used to transfer from any vessel at one stage to any other vessel at the subsequent stage. Between transfers, the line is kept under steam pressure. Transfer lines are sloped to drain with steam entering at high points and traps present at low points. This is universally true for all sterile transfers or sterile feed lines. Instrumentation should be sufficient to monitor and control seed development. DO measurements, for example, may not be required routinely in seed tanks; however, there should be ports available for special measurements during process development.

C. Medium Preparation

Since media prep is apparently a simple matter, batching and medium preparation are considered as an afterthought, if at all. Space is always at a premium and the supposed simplicity of this specific operation supports cursory planning that goes into raw material storage, transport, and *controlled* batching. Furthermore, planned inflexibility is dictated by a batching area designed for one or two processes. It is not often that the media make-up area of a plant matches the fermentation deck in design effort or appearance. Reference should be made to extent and accessibility of controls, cleanliness, lighting, and monitoring of performance. Media preparation design should involve:

1. Sufficient storage of wet and dry materials used as adjuvants or in relatively small quantity
2. Suitable flow monitoring and recording (flow or weight) devices for all ingredients
3. Well-agitated vessels with strain gauges or, at the least, clear and conspicuous volume markings
4. A clean area for laboratory measurements on media (solids, pH, clarity, refractive index)
5. Refrigerated storage of samples of media prepared (retain samples)
6. Temperature monitoring whether or not changes are programmed
7. Timely transfer (proper pump and piping design)
8. Bottom discharge vessels with methods for easy cleaning
9. Suitable disposal of empty bags or drums
10. Suitable scales (easily cleaned) for weighing smaller quantities
11. Safety features to prevent strains or exposure to hazardous materials
12. Adequate ventilation and lighting
13. Good communications to the rest of the plant

This is not meant to be an exhaustive list; it does point out certain potential design weaknesses. The media make-up area of a fermentation plant is not the first area shown to a visitor, if it is shown at all. Usually, there is a negative reason for by-passing the area even if secrecy is not at issue.

A number of articles and reviews are available which discuss fermentation plant design. Soderberg gives an overview of the entire fermentation plant, including auxiliaries and space requirements.[46] Plant layout is stressed and sterilization considerations are emphasized. Potential problems in control of sterility are well described. Charles[47] offers a critique of scale-up methods with an analysis of the correlations that have been published. Other papers discuss modeling, critical items to cover, considerations of scale, and generalized cost estimates.[48,49] A review of many patents is available in book form.[50] All products covered are prepared by fermentation and the reader will have available a large number of process flow diagrams for synthesis and isolation.

D. Fermentation and Harvest

The fermentor is obviously a key piece of equipment. All conditions within the vessel must be optimized for growth and biosynthesis. Sterility must be maintained, oxygen transfer and distribution must be achieved, all important parameters must be monitored or controlled, foam must be controlled, and streams must enter (and sometimes exit) in a controlled fashion. The design package must accommodate all these, sometimes conflicting, requirements. Most often the design package will present a compromise among the demands. Accommodation to all requirements might be possible, but the design would probably be extremely costly. This is most clearly seen in selection of the agitation-aeration combination. Scale-up on the basis of a single parameter is difficult because of concurrent changes in other variables. Simple scale-up based upon power/unit volume might require a vast power train plus expensive support structures. The ultimate selection does not satisfy all scale-up requirements, but the compromises are usually understood and are acceptable.

Table 13 lists desirable characteristics for pilot fermentors. The requirements for ultimate in flexibility and ease of handling are probably too severe for a production vessel; however, it would be better to eliminate a requirement because of potential overdesign rather than neglect that need in full-scale design. Therefore, the requirements should be listed and, once again, compromises made based upon the actual production needs contemplated and cost considerations.

Gas discharge is one important design item. Foam should be collected and possibly recycled (maintaining sterility) and exhaust gases may require special treatment depending upon composition or odor. Containment requirements may exist; for certain cultures, incineration of off-gas is essential.

Harvest requirements depend upon rate of downstream processing, need for speedy turnaround, pretreatment requirements, and stability of the final broth. Use of the fermentor as a harvest tank is wasteful of expensive equipment and will mean that plant throughput will be less than optimal. Very often, an atmospheric tank will be suitable as a harvest tank and as the feed tank for the first stage of extraction operations. The design will provide for periodic cleaning/inspection of the harvest tank; an in-place cleaning system is often justified. Material of construction is important, as everywhere in the plant, and resistant linings should be considered to save capital. Any gaseous effluents from the harvest tank should be controlled with suitable equipment. Monitoring (pH, temperature, viability, etc.) will be a function of lab and pilot evaluations.

The overall design package should be viewed from one other perspective. Many actual and contemplated products of biotechnology exhibit a relatively small annual *mass* requirement. Unit dosage may be extremely small. It is likely that a relatively small fermentor could supply enough broth to give an annual requirement for a single product. It would not seem an economic plan to continually add small fermentors and small extraction units as new products are developed. Some flexibility in design of a natural products plant is required. The perspective of flexibility will give the designer some leeway in equipment selection, spacing, and interconnection. In this area of biosynthesis and purification, it often happens that different products can be made in very similar equipment. Advantages and disadvantages of different approaches in plant design are given in Table 14. There are clear benefits in having a multifunctional plant; however, the costs (both qualitative and quantitative) should be considered as well.

After the drawings are prepared and specifications detailed it is a good idea to be certain that areas of major concern have been recognized. That is to say, not every issue must be resolved with no margin for error, but every important issue should be reviewed so that potential for major oversight is nonexistent. In short, the questions should be of the form, "What are the important design parameters and how have we tried to respond to the issues in this design?" The answers to this question will set the direction for additional laboratory

Table 13
DESIRABLE CHARACTERISTICS OF LABORATORY OR PILOT PLANT FERMENTORS

- Easily installed, readily movable, simple maintenance and cleaning
- Double mechanical seal, top or bottom entry, variable speed
- Complete piping system (pressure regulators, valves, filters, relief devices) integrated within unit frame
- Computer-controlled operation (display, disk drive, printer, keyboard), multicolor plotter a useful accessory
- Multiple ports in head
- Digital set point control with clear indication of setting, stable control, manual override provided
- Contact surfaces: stainless steel, glass, silicon rubber, Teflon®, Viton®, EPDM, standard gaskets, and O rings
- Nutrient additions at controlled, variable rate, programmable
- In-place sterilization, programmable preferred, filtered steam
- Heating or cooling for temperature control
- Aeration via sparger or draft tube, air outlet filter (condenser)
- Sterilizable sampling valve
- Ease of inoculation and sterilizable inoculum port
- Foam indication and control (include mechanical system)
- Separate control modules: temperature, pH, airflow, DO_2, pressure, nutrient addition, defoamer, level, pCO_2, redox
- Parameters programmable
- Multigas addition via sterile filters (with controls)
- Humidification of input gases
- Instantaneous or demand display: temperature, RPM, pH, airflow, DO_2, pressure
- Totalizer for nutrient addition, defoamer addition, acid, base
- Multipoint recorder (at least six variables)
- Expandable control modules and recorder
- Modular, plug-in systems for change or expansion
- Local or remote controls
- Alarm capability for selected variables
- Simple digital/analog conversion, relays for pumps and valves
- Batch, continuous or fed-batch culture
- Changeable impellers, removable baffles
- Properly coded
- Internal welds ground smooth, interior mirror finish
- Sterilizable drain (harvest) valve
- Jacketed (vessel heating or sterilization with minimal dilution)
- Aqueous or organic (as hydrocarbons) media
- Simple conversion to plant/animal tissue culture
- Breakers, fuses, overload protection
- Minimum requirement for spare or specially designed parts
- Quiet and cost effective
- Long-term, stable operation at temperatures of 15—50°C
- Capable of measuring cooling requirement (energy balances)
- Mass determination on vessel, feed, acid, base vessels
- Modeling to control measured variables (set point or algorithm) such as DO_2 via air or impeller speed or both or substrate concentration via nutrient feed rate
- Completely leak tested (timed hold of pressure or vacuum)
- Control valves fail to open or closed in logical fashion based on service
- Visible schematic slowing valves and measurement points
- Agitator power input by strain gage
- Sight glass or viewing port with illumination
- By-pass systems on key services
- Electrical connections and tie-ins unaffected by humidity
- Exit gas analysis via mass spec for O_2, CO_2, volatiles — linked to computer for control/monitoring
- Long-life sterilizable sensors (especially pH probe)
- Rapid heat up and cool down
- Minimum supervisory or manual interaction
- Automatic data collection and storage

Table 14
APPROACHES TO DESIGN OF A NATURAL PRODUCTS PROCESS PLANT

Dedicated	Generalized (multifunctional)
Advantages	
Sized to fit needs	Redesign, retrofitting minimized
Longer production runs	No delay in new process implementation
Reduced training needs	Quality control (hazard analysis) simplified
Little or no in-process build-up	Alternative paths easy to test
Lowest per-unit cost	Standby units may be used in emergency
Ease of training	Broader optimization potential

Table 14 (continued)
APPROACHES TO DESIGN OF A NATURAL PRODUCTS PROCESS PLANT

Dedicated	Generalized (multifunctional)

Advantages

Known utility and manpower needs	Adaptable in both scale and type of product
Amenable to computer control	Applicable to nonstandard or low volume runs
Lower insurance and taxes	Subcontracting or leasing is possible (custom manufacture)

Disadvantages

Inflexibility	Higher initial cost — potential overdesign
Difficult changeovers	Greater maintenance complexity
Expansions or major revision needed to accommodate volume change	Higher inventory of spare parts
Difficult to respond to change in demand or experimentation	Work-in-process build-up likely
	Long training periods/complex instructions
	Process planning may be more complex
	Complex instrumentation
	Utility and waste treatment loads may vary
	Greater need for warehousing
	Gross contamination potential exists

Table 15
DESIGN PARAMETERS — INDUSTRIAL FERMENTATION

	Variable or parameter	Result or analysis
Microorganism	Stability	Handling, rate of degeneration
	Nutritional requirements	Optima for growth and product synthesis
	Growth rate	Design of process
	Preservation methods	Maintenance of culture
	pH, temperature optima	Heat load
	Morphology	Agitation needs (power input)
	Response to O_2, CO_2	Aeration-agitation needs
	By-product formation	Recovery methods and cost, yield loss
	Ease of separation	Recovery method and effluent treatment
	Environmental or health aspects	Type of plant design (odor control)
Raw materials	Cost and availability	Costing or process, contractual requirements
	Alternates	Interruption of supply, cost cycles
	Uniformity	Response of process (yield, rate)
	Contaminants	Response of process (yield, rate, pretreatment)
	Stability, inventory	Storage conditions and turnover
	Ease of transfer (monitoring)	Energy needs, pipe sizing, instrumentation
	Ease of sterilization	Presence of particulates
	Effluent(s) composition	Cost of treatment or use as agricultural product
	Water quality	Need for treatment and supply problems
Process design	Number of stages	Sizing for inoculum development
	Cycle times	Design of plant, size, and layout
	Power requirement	Vessel and agitator design, variable power
	Cooling requirement	Vessel, heat transfer method, cooling method
	Air throughput (heat, pressure)	Compressor design and number
	Sterilization method	Continuous or batch, overdesign
	Nutrient feeding	Continuous or batch, monitoring
	Cleanliness	Sterility, clean-in-place systems
	Control scheme	Computerization, redundancy, local/remote

Table 15 (continued)
DESIGN PARAMETERS — INDUSTRIAL FERMENTATION

	Variable or parameter	Result or analysis
	Recovery options	Economic design
	Utility needs	Costs, cogeneration, alternate fuels
	Size of equipment	Cost and design of plant
	Maintenance	Percent on-time, cost, spare parts
Product	Stability	Type of extraction processing
	Response to upsets	Sophistication of controls
	Response to pH, temperature	Hold-up times, degree of control
	Purity (quality)	Extent of treatments, regulatory matters
	Environmental impact	Agricultural products, treatment cost
	Recovery options	Emphasize broth treatment, early separation
	Inventory	Special needs, refrigeration, segregation
	Transportation	Special needs, regulatory requirements

or pilot plant work, needed optimization, or further clarification by groups within or outside the company. One would hope that major design changes would not be needed at this point in time, since *any* change now has a cost attached to it. It is always better to correct a problem in design at this stage even at added expense; correction after the plant is built is far more costly. A list of key design parameters will, of necessity, be somewhat subjective. A general list is also less useful than a list prepared specifically for a given process and product. With this in mind, Table 15 lists variables or parameters that should be addressed during and after the process design phase. Results should be quantitative with error functions, but not all parameters are suitable for "hard" analysis. Qualitative information, even various alternatives, should be included as products of the analysis. The process design should be responsive to parameters of importance that have been recognized during the research, development, and scale-up phases.

E. Downstream Processing

Downstream processing is discussed in greater detail in a later section. With each passing month, a novel or modified procedure is described for selective separation. While more complex molecules in ever more dilute solution require more expensive and time-consuming processing steps, one should not overlook the most obvious purification operation, i.e., direct, one-step recovery. SCP is often centrifuged, washed, and dried. Solvents might be distilled right from the fermentor. While a unique fit is required, direct isolation or recovery should be considered first. Even if "impossible", one might eliminate a number of steps by creative flow charting.

In general, the usual unit operations are

- Broth conditioning (surfactants, flocculants, filteraid, heat, pH change)
- Centrifugation
- Cell disruption
- Filtration
- Absorption or adsorption (including ion exchange, gel filtration, chromatography)
- Solvent extraction
- Concentration (ultrafiltration, reverse osmosis, liquid membranes as well as conventional)
- Distillation
- Precipitation or crystallization
- Drying and classification

All the same concerns listed for every prior process step apply to downstream processing. The process section and design are developed and piloted so that a reproducible process results. Many trials are run, usually under near-optimum conditions. It is useful to have an analysis that covers suboptimal operations. If temperature control fails at one step, what is the best process alternative? What are minimum and maximum pH values for key steps that will allow continued processing? What is the response of the process to upset? What is the best human response to each upset? It is neither possible nor cost effective to cover every contingency. Still, upsets (planned or unplanned) do occur in lab or pilot plant practice. Responses to these upsets should be included in the design package; very often, these details are as useful as descriptions of optimal performance.

F. Waste Disposal

The changing regulatory environment, coupled with everincreasing treatment costs, creates serious problems in design of this section of the plant. Use of nearby municipal or county treatment plants is becoming more and more difficult as is finding acceptable sites for disposal of liquid or solid wastes. The criteria to follow here are, first, have a backup program in place (even if more costly) and, secondly, plan for expansion.

Every side stream must be considered a waste stream. As such, opportunities do exist in the area of animal feed supplements. If bioactive compounds are present, a careful check is needed to determine allowable use or levels of use. Selection of inputs (including defoamers, processing aids, and every other stream entering the process) may change a problem to an opportunity. An oversight at the design phase concerns disposal of a contaminated batch. What are alternatives as a function of the contaminant? Are shock loadings a problem to the treatment plant, on-site or off? Can a contaminated batch be processed, reprocessed, or converted to something marketable? These and similar questions will have been reviewed and reasonably firm answers should be recorded. A design error at this point may result, not in a 1- or 2-day shutdown, but a lengthy and costly shutdown while major equipment installation is made to correct a planning flaw. Some sense of future requirements (meaning regulatory limits) is needed to plan appropriately. All licenses and permits must be obtained in early design and construction stages. This could involve local, state, and federal authorities and could take many months or years.

G. Final Forms

It should be clear to the planners that isolation of a very pure, biodynamic compound via fermentation at a commercial scale does not end the production cycle. In fact, some might say that the work really begins at this point. Even if complete quality specifications for the bulk material were detailed (and they should be) and satisfied, there remains a host of design factors to consider. In a multiproduct plant, potential for cross-contamination is ever present. Antibiotic cross-contamination is merely one example. Where and how should the bulk product be stored and shipped? What is shelf life and what are degradation products? Are there hazards to workers (not only immediate but due to lengthy exposure)? What are final dosage forms and how are these to be made and stored? What testing occurs with aging? What stable tracers can be added to identify the product? There are many pharmacology questions which must be answered but which are beyond the scope of this book. The designer, however, must have appropriate input to establish physical needs and associated investment and operating costs.

Final product specifications may result from laboratory and pilot runs. To affect the same product at a plant scale might entail a massive incremental investment. Finished product quality and efficacy must be assured; however, there should be some relationship to what is feasible and practical.

H. Nonprocess Plant Design: Process Documentation

Documentation would appear to be a subject that is not complex and readily satisfied. In a project of any magnitude, such is not the case. Documentation is usually taken to refer to laboratory and pilot data that are the basis for the design; this is only a small part of the total. Not only should primary data be stored correctly, but every basis for a design decision or, more critically, a design change should be easily available.

Process documentation continues to flow "to" project management for the life of the project. Even after successful start-up, process and equipment changes continue. Useful documentation means that errors will not be repeated; i.e., history does not have to be relearned on a periodic basis. It is best to begin segmentation of documents very early in the project cycle. Natural divisions are obvious and no less useful for their simplicity. Some major headings are

Laboratory data	Flowsheets
Pilot data/process calculations	Equipment selection/specification
Quality control tests	Bids and estimates/guarantees
Hygiene/toxicology data	Maintenance details
Planning/scheduling	Instrument details
Manpower requirements	Operating instructions
Job descriptions	Vendor bulletins/drawings
Start-up plan	Safety requirements/hazards
Training/emergency procedures	Contracts
Test (in-process) methods	Detail drawings

Design change orders

Responsible personnel (who change in the course of the project) should have convenient access to all these data; proper indexing and organization of the files are essential. Engineering drawings should be kept in a neat file that inhibits degradation (or mishandling) while allowing review without unfolding tens or hundreds of drawings. The drawings should encompass "as built" construction, plot plans, process and instrumentation drawings, iso-metrics, undergound lines and all schematics (electrical and instrument). A letter file and monthly reports (pre- and postconstruction, pre- and poststart-up) should be in the file. Calibration results on instruments and vessels should be available. (Regulatory requirements have a major impact on what is done as well as frequency.) Any special tests (environmental hazards, coding, leak tests, radiographic tests) should be maintained in the file.

The advent of computer files is a boon. However, one could imagine the result of an inadvertent erasure. Redundancy is important, whatever the file system. The success or failure of a documentation program is usually apparent after the fact. In other words, the "missing" document or piece of data is only known to be absent when the need is greatest; this may occur years after process start-up. The best guide for a properly executed docu-mentation program is a successful project with information availability for many years afterward, that is, a successful filing and retention program. This may not be available for a like project, but similar programs and needs exist even for building a general services building or a major process modification. Should a model not exist, it is probably better to retain more data and purchase information than is optimal. Organization may be more complex and storage space will be greater than optimal, but the additional cost and effort will be minimal compared to cost related to absence of needed data.

I. Utilities

Water supply is often a very important consideration in site selection. Fermentation processes use very large quantities of water for media preparation, steam generation, ex-traction operations, washing, and cooling. Composition for each intended use is different;

pretreatment is often essential. If there are seasonal fluctuations, storage should be considered. Backup supplies should always be "designed into" the process or site so that interruption from one source does not shut down the plant.

Other decisions must be made prior to design. Which forms of energy should be purchased? Which self-generated? It is usually a good idea to have multiple sources of supply for purchased energy and if a fuel is to be burned, multiple sources should be usable; that is, any two of gas, liquid fuel, or coal should be sufficient; one can be stored in sufficient volume to compensate for interruptions. Furthermore, the ability to use multiple fuels gives the company an opportunity to establish an energy equivalency minimum cost for the main fuel supply. Cogeneration must be considered at the preliminary design phases; fermentation operations are ideally suited for cogeneration systems and attendant savings.

In process design of utilities, a conscientious effort must be made to minimize capital investment. The development and scale-up stages should have included studies related to reduction of energy costs. Should these studies be lacking or deficient, some directed experiments can be run to establish alternate, less costly methods of supplying energy. Supplying energy is meant in the broadest sense; i.e., any energy that must be expended to satisfy process steps is included. Some examples of energy optimization are

1.   Minimum cost combination of agitation and aeration (discussed earlier)
2.   In-cycle changes in airflow and/or agitation to satisfy oxygen demand while lowering energy costs
3.   Economic insulation
4.   Condensate recycle
5.   Heat recovery wherever feasible (continuous sterilization, boiler exhaust gases, carbon regeneration exhaust)
6.   Use of cooling towers, wells as opposed to mechanical refrigeration
7.   Economic determinant for drivers (turbines vs. electrical)
8.   Mechanical vapor recompression or steam jets on evaporators
9.   Air removal vs. steam displacement of air prior to vessel sterilization
10.   Cogeneration
11.   Liquid ammonia (neutralization) in or at fermentor to gain heat of evaporation

These points are not applicable only to a grassroots plant; retrofitting or process optimization can and should include these considerations. In fitting a process to an existing plant, one will normally find that an excess of water, steam, or energy is not always available. Reserve capacity has a habit of being rapidly depleted. It may be necessary to evaluate a process at suboptimal conditions in a retrofit situation. Very often, the reduction in rate or yield is not great. Not only is a major capital expenditure avoided, but there is a great savings in time.

Selection of drain systems (storm sewers, sanitary systems, plant drains) may seem a mundane task. However, a knowledge of local and federal regulations coupled with appropriate siting of plant facilities and storage will result in a design that is not only capital cost effective, but will result in ongoing operational savings. Only liquid streams that require special treatment should be diverted to that operation; varyings levels of treatment may be appropriate with only those streams requiring complete or complex treatment so collected. It makes no sense to send a high volume stream that requires only pH adjustment or settling to an aerobic digestion system. Further, spill protection must be designed so that appropriate drainage or collection occurs. A massive spill of strong acid or base to the storm sewer system is not a desirable design consequence.

J. Other Buildings

There are many auxiliary functions that require planning and design. These are not directly related to the process. In a major production plant, an added process or even a new production building may not require major additions to auxiliary services. For example, an existing quality control laboratory (with instrument and/or personnel additions) may serve both existing and new processes. Unless massive product volumes result, a completely new warehouse is seldom indicated. In a grassroots plant, however, all these considerations (and others) must be reviewed in a careful and detailed manner. The problems and costs associated with underdesign or overdesign are obvious.

Major space requirements and need for proper siting exist for:

Personnel facilities
    Male and female changing rooms, personnel lockers, clothing changes
    Storage of personal protective equipment
    Elective or mandatory showers
    Access to parking
    Security needs (especially on off shifts, weekends)
    Locations for eating, smoking
Administrative facilities
    Plant security and safety
    Medical and emergency treatment
    Communications
    Computer network (in-plant and external)
    Support laboratories (development, tech services)
    Cafeteria
    Documentation, design, and start-up files
Maintenance
    Locations for crafts (central or local)
    Spare parts (indoor and outdoor storage)
    Planning and scheduling (work order system)
    Receiving, security, check-out procedure
    Lubricant storage and disposal
    On-site equipment needs vs. lease/rental
    Instrument/electrical calibration
Quality control
    Instrumentation needs (temperature/humidity needs)
    Regulatory requirements (computer storage)
    Adaptability to new processes
    In-process and finished goods control, coding
    Label files and security
    Storage of specialty gases and standards
    Retain samples
Warehousing
    Traffic patterns in and out
    Security and inventory control
    In-process storage, rework, isolation areas
    Degree of automation
    Special temperature, humidity requirements
    Protection against infestations

The listing is by no means exhaustive. The usual concerns of energy conservation, fire protection, emergency procedures, access, internal and external communication, lighting, roads and walkways, and upkeep relate to every structure. The intangibles of morale and elan are very often strongly related to the aesthetics and efficiency of the "auxiliary" structures and services. The designer neglects these factors at his own, often considerable, risk.

A major result of the process design phase is a process flow chart. This chart may be in the form of pictorial representation (unit operations and unit processes) or in the form of box diagram. Essential information on each step of the operation is listed with references to more detailed charts or tables. Sizing of equipment may be implied or shown, but the main objective is to present the necessary sequence of steps with inputs and effluents that are essential to affect synthesis and purification. Energy and material balances can be shown later, but flow indications direct the effort to further detail or magnification of each process step. Two alternate methods are shown for citric acid (Figures 4 and 5). Figure 4 is a flowchart for citric acid production. If substrate pretreatment is required, it, too, should be included. Some greater detail might be included, but detailed piping and electrical drawings are not normally available at this point in time. Figure 5 emphasizes crude and final purification of citric acid. Certain volume and yield data are shown; starting point is harvested whole broth. The patent literature is useful for flow charting; even if an exact scheme is not available, some similar categories can be screened for useful methods. The flow chart is used for scoping plant scale, equipment selection, equipment siting, as well as giving the finished process outline, albeit in abbreviated form.

45

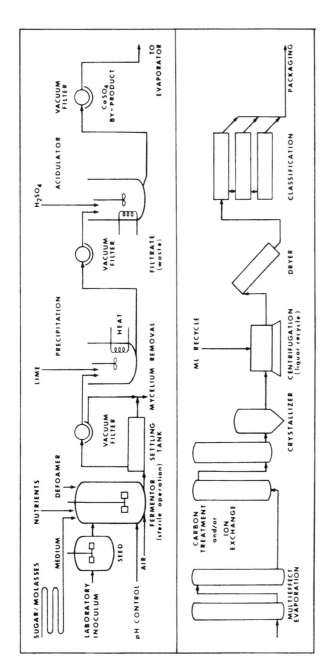

FIGURE 4. Flow chart for citric acid production.

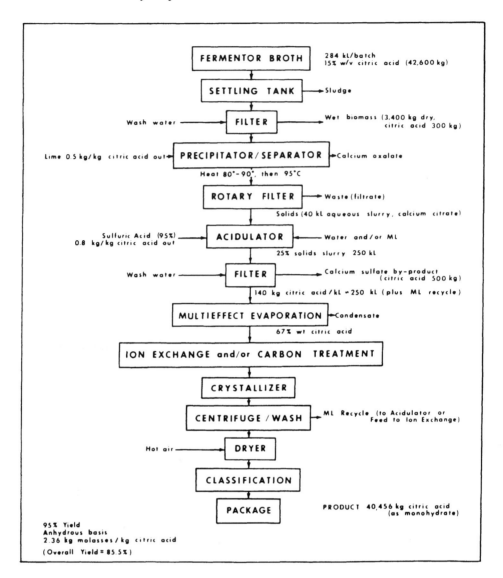

FIGURE 5.   Citric acid purification.

Chapter 2

ENGINEERING DESIGN

I. INTRODUCTION

Once a decision is made that a potential need exists that can be fulfilled by mass production and, furthermore, that the production can be scaled up, some rather weighty questions arise. It is probably better to say that these questions have been known since the inception of the most fundamental research, but there has been no need to address them until now. Certain of these questions overlap the process design phase:

1. Should the production occur in an existing plant, a retrofit facility, a leased plant, or in a grassroots plant?
2. How much product is wanted? Should there be expansion capability? What raw material and product storage is required?
3. Is time of the essence? How much is it worth to complete the plant 1 or 6 months early?
4. Should design be single-product oriented or is a fair amount of flexibility to be "designed in"? Should planners, in other words, anticipate other generations of products?
5. How much redundancy is needed? The converse is how many single train, critical pieces of equipment can be permitted?
6. What is the degree of environmental and regulatory requirements for various sites and for different products contemplated?

These and other questions are focused upon the fundamental issue: for the investment in time, money, manpower made, what is the expected return that can be reasonably expected? The engineering design staff has as its focus (1) the establishment of a highly accurate investment estimate and (2) the creation of controls to insure project completion within preestablished economic constraints.

An important element in design concerns alternative approaches. There is an inherent risk present at all stages in the project cycle; risk is magnified in a rapidly evolving, highly regulated industry. Fermentation and biotechnology fall into this category. Every project involves some level of assumption (all essential data on production and scale-up are never available, to say nothing of sales and marketing projections) and any assumption has an ongoing risk associated with it. Since risk is unavoidable and it is hazardous to simply disregard it, measures of risk analysis have been developed and varying degrees of sophistication employed to at least define risk, since it cannot be eliminated.[51] In uncertainty analysis, an item, calculation, projection, or alternative is identified and probabilities associated with each such term or path. One end result is a series of alternatives or a plot of a response vs. a probability term. Two or more alternatives may be compared in a quantitative fashion. Although more complex than a one-column table giving a single quantity for each variable, this method presents ranges of assumptions with differing risks and responses. Ultimate selections, hopefully, will be based on a broader and more reasoned input.

At some point in time near completion of the final research document, or just afterward, a request for funds is prepared. Each organization has its own format and name, but the title — whether capital request, authorization request, or investment authorization — clearly indicates that the subject document embodies a request for funds along with complete justification for such investment. It is the usual case that a low-end cutoff for funds exists below which a simple one- or two-sheet form is sufficient. There is normally no high end limit; however, as a general rule, documentation and supporting evidence are present in direct proportion to the dollar amount requested.

Listed below is information required for a capital request. Extent of detail is dependent upon the company philosophy and variations in format exist. The list does give a reasonable outline and is a good starting point.

Objective and rationale
    Confidence level, both technology and market
    Reason for selection of this alternative
    Justification for proposed project, especially corporate "fit"
Marketing information
    Analysis with projection
    Market test data, efficacy, pitfalls
    Competitive analysis
Process and technology
    Patent or proprietary position
    Laboratory and pilot data, reproducibility
    Detail of proposed project
    Fit to existing plant (if applicable)
    Environmental requirements and potential problems
Safety and loss (special testing requirements)
Long-range investment
Responsibility assignments
Additional demand on existing plant
    Manpower
    Utilities including waste treatment
Risks (schedule, environmental, regulatory, cost, price, volume, process, obsolescense)
Corporate approvals
Start-up expenses
Schedule
Reliability of estimate (plus explanation of contingency)
Staffing requirements
Financial summary
    Capital and expensed costs
    Net cash outlay
    Change in working capital (inventories, receivables)
    ROI and cash flow
    Payback period
    Escalations
Alternatives (including "do nothing")
Sensitivity analysis
    Selling price
    Volume
    Costs
    Capital estimate

Once a decision is made to evaluate an investment (note that the decision to invest is deferred), some estimate of required expenditure is needed. Accuracy of an estimate is dependent upon:

1.   Period in project cycle when estimate made (the earlier, the less accurate; however, the earlier the better)

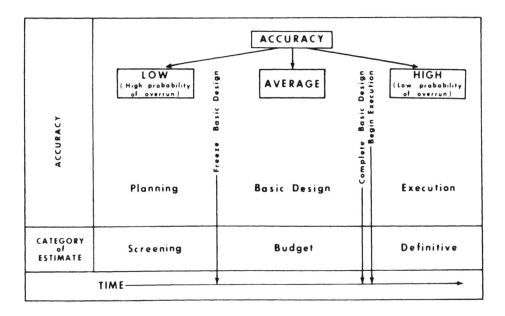

FIGURE 6. Estimate accuracy as a function of time.

2. Prior history of estimation for like processes and products
3. Basis (assumptions) of the estimate
4. Corporate approach to project evaluation and execution
5. Skill of the estimator(s)
6. Completion of research and process development phases with sufficient documentation

There is a well-known relationship between estimate accuracy and time (see Figure 6).

The methodology and techniques used in cost estimating are detailed in a number of texts. One gives a stepwise approach based on phases of the project cycle.[52] The direct costs involve material items incorporated in the plant plus related materials to effect installation. Labor, then, is a direct cost. There are a number of indirect costs which must be incurred and methods of estimating them are also given. Indirect costs are subject to a higher degree of variation; these costs include:

- Detailed engineering
- Contractors' fees
- Temporary construction facilities
- Construction consumables/tools
- Field supervision
- Labor payroll burden
- Insurance, freight, duties, taxes, permits
- Owner costs

Careful detail is required in preparing an overall estimate; methods are described for estimate coding and estimate documentation. There is a thorough review of contingency and the need for a careful *estimate* of contingency plus the need for careful control of this large grouping of funds. Elements of control of contingency funds are

1. Document basis for contingency estimate — this includes specific definition of items included

2.   Educate all participants in how and why contingency derived
3.   Control changes either up or down; contingency is not a floating quantity
4.   Reestimate on a routine basis (as monthly)
5.   Contingency is not a "slop fund" or a buffer factor to accept design changes or favorite projects that are not part of original scope

Project management responsibilities are centered around:

- Quality: plant designed and constructed to owner's standards; plant to operate safely for specified period producing quality product
- Schedule: complete and start-up plant at prespecified time
- Cost: complete plant within specified budget so that financial predictors of return can be realized

Cost-control strategies, relative to the overall cycle, are described. At each phase of the project, there are different techniques to keep costs under control and to react quickly when reality moves away from expectation.

   The subjects covered in the design-construction cycle, as well as relative timing, are given below. These factors must be considered and quantified in a sequenced and controlled manner.

Evaluation

   Early phase
        Alternate processes
        Alternate yield studies (varying degree of rework/rejection)
        Differing percentage of on-stream time
        Alternate locations
        Order of magnitude (scope)

   Basic design
        Selected process (minimal future modification)
        Location of facility and type of plant
        Utilities and off sites
        Preliminary PERT chart (timing of project)
        Potential contractors/constructors
        Conceptual estimate ("budget" estimate)
        Initiate documentation

   Final design
        Process is "frozen" — design changes will be costly and
          will probably involve schedule delay
        Final detailed design (heat, mass, component balances)
        Selection of contractor
        Detailed engineering and final utilities requirements
        Initiation of stringent cost controls (control of changes)
        Detail drawings
        Control documentation (organize manuals)
        Economic sensitivity analysis (cash flow estimate)
        Definitive estimate prepared
        Environmental controls (discharges) fixed

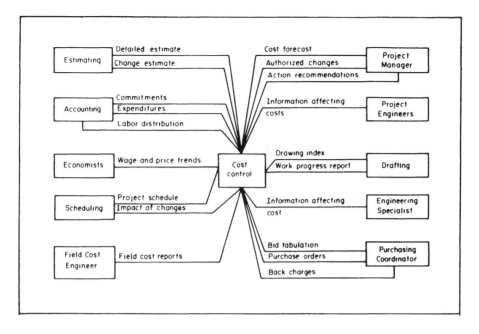

FIGURE 7. Cost control flow chart. (Reprinted from Clark, F. D. and Lorenzoni, A. B., "Applied Cost Engineering," pg. 151 by courtesy of Marcel Dekker, Inc.)

> Construction
> Bids and contracts
> Planning and scheduling
> Field engineer and field team
> Control of direct labor
> Subcontracts
> Guards (site control)
> Documentation of expenditures (code of accounts)
> Monitoring of progress (percentage completion)
> Control of overtime and change orders
> Miscellaneous overheads
> Inspection

A cost control flow chart shows the flow of information both in and out. Any interruption in timely information transfer will cause dislocations and errors. Not only will the estimate be off, but final costs will be off and design operability may not be achieved. The actual elements of cost control and timing will be discussed later. Figure 7 shows the many relationships that exist in design and execution. Not every title (specialist) is required on a small- or modest-scale project; however, as scale or cost rises, it is strongly suggested that the various inputs are established. If necessary, a consultant or outside firm might supply needed data. If a contractor is used, certain functions can be assigned to that organization. Successful cost control depends upon continuous and meaningful contact between that person having the assignment and the necessary auxiliary functions that monitor planning and expenditures.

With this general introduction complete, we can move on to design problems and approaches in fermentation processing.

II. REACTOR DESIGN

Since the heart of a fermentation facility is the bioreactor, a significant design emphasis is needed. It is rather remarkable that in the approximate four-decade history of large volume aseptic fermentation, no design has supplanted the stirred tank reactor (STR). The post-World War II antibiotic fermentations required voluminous air input, medium and feed sterility, good mixing, and careful process control. Further, corrosion could not be tolerated (cation contamination was recognized as a potential problem early on). Stainless or stainless-clad pressure vessels with a sealed agitator shaft and with sterile air input were used in the mid-1940s and some of those units and their more up-to-date replacements are still in use. Improvements have been many. In the early days, total vessel volume was often 10 to 20,000 gal; today, volumes five- to tenfold greater are in use in a single STR.

Fermentor design should include these reviews and concepts:

1. Usually, more than one fermentor is required. Plant layout (plan and elevation) should be planned early.
2. Access floors (sampling, maintenance) must be provided. Good lighting should be planned.
3. A fermentor deck is usually humid (steam leakage, wash water, condensation). Provision for drainage and humidity control should be made.
4. Piping runs are long and complex. Extra effort in planning will prevent maze formation.
5. Wiring for conventional and computer control is extensive; early planning is essential.
6. Location indoors or out is permissable. A risk-benefit analysis is warranted.
7. Noise, odor, environmental issues should be reviewed early. A fermentor deck is noisy and large volumes of exhaust gases are very often involved.
8. Cycle time should include charging *and* harvesting. Feasibility and cost must be covered to settle upon acceptable turnarounds.
9. Seed vessels, auxiliaries, sterilization systems, feed tanks, and defoamer systems must be included in layout.
10. Value of modeling should be considered. In a complex, multifunctional plant, it is strongly suggested.

Scale-up considerations as well as alternate configurations are discussed in review articles.[47,48,53,54] Mass transfer and turnover considerations are also included. The use of macroscopic methods, i.e., analysis of stoichiometry of growth and heat production, can be applied to fermentor design.[55] Empirical estimation of OTR can be related to overall energy balance and fairly narrow ranges set for reactor/agitator design.

There are specific details which should be incorporated in fermentor design (beyond shape, dimensions, and weight):

1. Number of inputs and outflows as well as location in vessel (outside location plus depth within fermentor)
2. Type of support structure (agitator independently supported?) for loaded vessel
3. Size and type of agitator, if any; alternatives are draft tubes or air agitation only
4. Pressure and vacuum rating and types of relief devices; example is operating pressure of 30 psig; design for 1 2/3 × OP
5. Type of seal on agitator shaft; top- or bottom-entry agitator
6. Heat transfer surfaces — type and quantity, internal or external; design temperature established
7. Amenable to fed-batch or continuous operation
8. Type of mechanical foam control — internal or external, broth recycle

9. Sampler design and location; consideration to continuous sampling of broth, exit gas
10. Valve types and locations, by-pass systems
11. Steam seals and traps
12. Access to internals — repair and inspection
13. Metal and weld finish — weld testing, leak testing
14. Type and location of controls and alarms
15. Central control room and local indication
16. Flexibility in changing impellers or impeller location
17. Mass/volume indication
18. Height/diameter ratio
19. Materials of construction, especially contact surfaces
20. Minimize longitudinal seams; flanges to have machine-finished faces
21. Detail stainless to stainless welds; detail stainless to carbon steel welds
22. Gasketing materials specified
23. Code specifications to be satisfied and so stamped; designed, fabricated, inspected, tested in accordance with ASME or API code
24. Shipping and fabrication details to be specified
25. Method of cleaning
26. Thermowells with welding details
27. Heat treatments to vessel should be detailed

It is probably best to have approved detail drawings for:

1. All nozzle details and locations, including thickness
2. Sight glass detail and location
3. Internal supports, especially for coils
4. External supports including clips, lugs, saddles, and orientation
5. Manhole entries and covers, all davits
6. All major welds, including overlap and offset
7. Entry and exit locations
8. Agitator support
9. Baffle details and location

If not on drawings, detail:

1. Maximum variation in diameter and straight shell
2. Maximum variation in bolt circle diameter
3. Orientation variations
4. Variation from centerline or reference plane to specific installed fitting or part
5. Tolerance allowable on parallel faces
6. Tolerance on perpendicular installations

While every vessel should have an inspection checklist, it is a good idea to inspect the fermentor as it is being constructed and erected. This means that the plate purchased for the project should be inspected prior to rolling. A limited inspection checklist for a fermentor should include:

1. Material identification
2. Review of (or presence at) pressure tests, stress-relieving tests and charts, radiographic analysis
3. Dimensional check

4.   Completeness of internal and external attachments and fittings plus location and orientation
5.   Check of finish
6.   Check of workmanship and cleanliness
7.   Witness hydrostatic tests and any special tests of pads or supports
8.   Check painting, if any
9.   Verify nameplate data

Sitting reviews many mechanical details as well as newer fermentor designs.[54] Some developments in construction have the potential for improving vessel integrity while lowering costs.[56] Brown[53] concludes that interacting constraints on vessel size result in a "common" volume of 125 m³ (4-m diameter by 10-m high) with upper limits at 210 m³ (4.5-m diameter by 14-m high).

III. AIR

Although air compression and supply is an auxiliary, in a strict sense, the importance of air to the fermentation process permits inclusion of this section at this point. It is desirable to have oil-free compressed air. For small fermentors or even a single unit, nonlubricated screw compressors (two or variable speed motors) can be considered. Centrifugal air compressors are available to deliver 150,000 ACFM at up to 5000 psig. Some redundancy is required both for preventative maintenance and mechanical failure.

Design of a delivery system should be based upon maximum requirements, not average. For multiple fermentors, proper scheduling will reduce peak demand. The design pressure of the delivery system depends upon these variables:

1.   Pressure drop in lines and valves
2.   Pressure drop in air filter
3.   Pressure drop in sparger
4.   Liquid head in fermentor
5.   Overpressure in fermentor

Air supplied to the fermentor must not only be liquid and oil free, but must be sterile. Use of adiabatic or quasi-adiabatic compression temperatures, with or without holding sections, will allow varying degrees of microbial kill. Discharge air will be pasteurized or sterilized depending upon compressor operating temperature. Discharge air must be at a pressure such that at the point of entry into the fermentor, pressure is 10 to 15 psi over that required by vessel pressure and liquid head. The air distribution system should be free of condensed water and a minimum temperature should be maintained throughout.

Wherever possible, lower pressure blowers should be used; they are much less costly to purchase, operate, and maintain. If liquid head and overpressure are both low, a blower may be suitable. Use of draft-tube systems may allow use of a blower. Pressure drop considerations become more important as discharge pressure is relatively low.

Air entering the compression train should be relatively free of dirt and larger particles; some coarse or primary filtration is suggested. Air intake should be located well above ground level (and well above roof level, if possible) and well away from any gaseous exhausts.

Air filtration has been achieved by various means; carbon and glass wool-packed beds have been successfully employed. More recently, thin-walled devices down to membrane systems have been installed and are in use. The support and submicron separators are steam sterilizable. Major cleaning effort or regeneration is not economic as replacement cost is relatively low. Useful life with relatively clean air is long and developments are extending utility even more. Air temperature, even with heat regeneration, is maintained so that relative humidity remains well below the dew point. This is true even with absolute membrane filters, since system design normally does not allow for significant water impingement.

There are a number of advantages to use of air only for agitation as well as aeration. Some of the advantages are

1. No need to support large motor and drive assembly plus shaft and impellers; vibration potential eliminated
2. Reduced overall maintenance
3. No need to have shaft entry; one less point to seal from microbial entry (or exit)
4. Heat of mechanical agitation eliminated
5. Shear sensitive cells can be grown more easily (mammalial or plant cells)
6. Higher air throughput means reduced CO_2 tension and reduced potential for inhibition
7. Power input and turnover varied easily

Use of air to mix fermentation broths is reviewed by Soderberg.[46] The characteristics of various compressor designs are given in a series of articles.[57] Limitations, efficiencies, maintenance requirements, and advantages for each type are detailed. Calculation procedures and typical examples are shown; investment economics (for different driver selections) are also detailed.

IV. DOWNSTREAM PROCESSING

A. Introduction

Biosynthesis is the critical first part in preparation of a product. In most processes, the greater portion of capital and operating expense relates to extraction and purification. The equipment is usually complex, many stages and different unit operations are involved, and reagent and labor costs are appreciable. In general, all the stages which follow the fermentation stage — starting with the harvest tank — are combined to be called "downstream processing".

A very simple downstream process would be (1) harvest, (2) concentrate, (3) dry and package. Such a process is rare (except for certain yeast and animal feed products) and relatively inexpensive. Other downstream processes are of such specificity that a "generalized" process cannot be detailed. Both the nature of the fermentation broth and the nature of the bioactive component dictate the purification. Factors which determine subsequent steps are detailed in Table 16. While many are obvious, one or more factors may be neglected in early development. An oversight at this stage could lead to process overdesign or to a scheme that is inoperable on a production scale.

Cost considerations must enter the planning cycle at the earliest stage. Very often, some quantity of finished product is wanted "immediately" for clearance work or to perform further use testing. Early process selection may be fixed prematurely, because the "same" product will be required at a later time. The opportunity for selection of the least cost option will be lost. Even if definitive costing is not possible, assumptions can be made which will assign the same margin of error to all choices. Costs will vary for all the well-known reasons:

● Production site is unknown.
● Taxation, insurance, depreciation are unknown.
● Knowledge and capabilities of workforce vary.
● Availability of reagents, utilities, services is unclear.

Table 16
PROCESS DETERMINANTS FOR DOWNSTREAM PROCESSING

Whole broth
    Nature of the bioactive product
        Intracellular
        Extracellular
        Lability
    Broth viscosity and emulsification tendency
    Foaminess
    Odoriferous components
    Biohazards
    Filterability
    Waste treatment, needs, and costs

Product
    Ionic or nonionic
    Solubility (polar and nonpolar solvents)
    Similarity to other media/broth constituents
    Purity
    Stability under processing conditions (temperature, pH)
    Yield losses
    Sensitivity to oxygen, other gases, or liquids
    Mol wt and conformation
    Hygroscopicity
    Allergenicity or antigenicity

While the overall scheme for any specific process cannot be detailed, some description of the various extraction operations is in order. A number of general articles can be consulted for an overview. Atkinson and Sainter point out that product recovery will become an increasingly dominant factor in overall production cost, especially for low volume specialties.[58] Many processes are categorized and processing alternatives given. Available unit operations are shown in typical sequences. Howell discusses individual steps in downstream processing from a scale-up and engineering perspective.[59] Unit operations are reviewed separately. A listing of extraction (henceforth, extraction will include purification, concentration, and other operations which can be pooled under a single heading) steps are given in Table 17. Consideration must be given to integration of the extraction process with improvements in fermentation. Examples are

- Recycle of cells to improve rate
- Continuous fermentation
- Eliminate need for filteraid supplementation
- Change in media to give useful (or at least nonhazardous) by-product streams
- Overdesign to handle higher product concentration in same broth volume
- Handling of contaminated or low-producing batches
- Extraction directly from the fermentor (as vacuum stripping)
- Recycle of valuable media constituents with or without partial purification

These items are self-explanatory. One, however, deserves some discussion. If an extraction operation includes ion exchange or chromatography, sizing to a specific product load, or even to 200% of design loading, may limit plant output unnecessarily. True, additional units might be installed in parallel at a later time, but this is costly and time consuming. If the fermentation process results in a low concentration of an active ingredient and there is some expectation of yield increase due to genetic manipulation, the extraction design should incorporate such expectation, within reason. If a 3-ft-diameter column is suitable, a 4-ft-

Table 17
DOWNSTREAM PROCESSING IN
FERMENTATION OPERATIONS

Centrifugation	Filtration
Settling (flocculation)	Cell rupture (disintegration)
Concentration (evaporation)	Solvent extraction (recovery)
Steam or gas stripping	Distillation
Precipitation	Crystallizaiton
Ion exchange	Adsorption
Ion exclusion	Chromatography (includes HPLC)
Carbon treatment	Affinity partition
Hydrolysis	Aqueous extraction
Chemical reaction	Affinity precipitation
Immobilization	Reverse osmosis
Dialysis (electrodialysis)	Liquid membrane
Preparation of dosage form	Ultrafiltration
Freeze drying	Gel filtration
Formulation	Washing (resolution, reslurry)
Racemization	Drying
Liquified gas extraction	Sterilization

diameter column of the same height would increase adsorbent volume by 178%. Installation cost would probably not rise by much. Other considerations (such as cost of adsorbent and cost of regeneration per cycle) must be included, but the advantages of improved throughput *immediately* after an important fermentation yield increase is attained should not be overlooked.

Another consideration not often evaluated early in development concerns economic balance between product accumulation rate and extraction operations. Computer simulation now simplifies a potentially tedious problem. Most product accumulation curves follow an S-shape. There is a decline in volumetric rate in the latter stages of fermentation. Most costs (air, cooling water, labor, overhead, etc.) do not decline, but are linear with time. Therefore, an economically optimum cycle time can be calculated once certain operating restraints are applied. As product concentration rises in a constant broth volume, unit extraction costs will decline for reasons much the same as those given above. These are diametrically opposed constraints and any change in fermentor cycle to improve economics must be analyzed in light of added or reduced costs in extraction. These analyses are not limited to the developmental phase; they are possible even during actual production even if ranges are somewhat more constrained.

In the area of medicinals and pharmaceuticals, total mass output is usually small in comparison to conventional industrial chemicals. Whereas fermentors, harvest tanks, and cell rupture devices can afford general applicability, more specialized, subsequent processing steps mean dedicated equipment must be installed. Any extraction equipment should have the greatest potential flexibility even if installed cost is somewhat higher. One may also conclude that the newer bioactive compounds are more complex, probably more labile, require greater impurity removal (or, conversely, must exhibit greater physiological purity), are present at lower concentrations in broth, and may command a much higher unit selling price. This, in turn, means that separations will be more difficult and costly. It has been concluded:[60]

Additionally, the market for these specialised materials is more concerned with quality than with cost, as the cost of the materials is generally only a minor component in the overall cost of the activity in which they are used (whether they are enzymes for an industrial process or medicinals). The emphasis on the recovery processes is thus on quality of product, yield and reliability rather than the cost of operations.

Items which must be included in costing of extraction operations (details are given in later sections on costing) follow:

Fixed capital	4—6 × cost of equipment
Depreciation	Straight line (11 or 12 year)
Labor costs	Geographic and process independent (range of $25—40,000/ man-year)
Raw materials	Process dependent (range 10—30% of product cost)
Overhead costs	5—25% of product cost
Percentage running time	Less than 100% (80—90% usual)
Cost of utilities (ranges)	Strongly a function of geographic location
Steam	$4—7/1000 lb
Electricity	$0.04—0.08/kW hr
Natural gas	$2.50—3.50/1000 CF
Cooling water	$1—10/1000 CF
Process water	$1—7/1000 CF
Fuel oil	$0.60—0.80/gal
Process air	$0.15—0.30/1000 CF

Some generalized methods can be used to estimate cost of downstream processing. The flowsheet can be segmented and cost per unit volume (or per unit product) can be related to each processing step with appropriate provision for recycle or side streams. The generalization can be very simple, meaning so many dollars per step (or per 1000 gal or per operating hour) with a simple multiplier for steps employed. There is no correction for energy use, actual labor needs, or even additional supplies. With sufficient historical information coupled with use of conventional technologies, a surprisingly good cost estimate results given the lack of effort or detailed calculation.

The next step in refinement would introduce a measure of detail, i.e., raw material usage, yield, heat (energy) balances, recovery of solvents, labor, and overhead costs. The measure of detail must be proportional to the detail of the process flowsheet. The hierarchy of detail must occur in all aspects of the design. In simple terms, data are no better than the underlying assumptions. At this level of detail, certain ranges or multipliers can still be used. However, knowledge of the process will limit the ranges used and will allow some reduction in percent error. If an existing facility is to be used and there is no capital investment or added personnel, cost allocation is relatively simple. The basis of allocation is the listing of current costs.

B. Specific Operations

A number of specific operations and associated costs will be discussed.

1. Broth Treatment

One effective technique, often overlooked, is activity within the fermentor or harvest tank to simplify subsequent operations and effect cost savings. In-fermentor examples are

1.   Dialysis or UF systems within the vessel
2.   Suspension media (tissue culture) with reuse
3.   Vacuum stripping
4.   Formation and removal of precipitates and complexes

Second, a harvest tank can be used to the same effect allowing reuse of the fermentor. Other treatments might be flocculation, pH change, or lysis.

These processes are geared toward major savings in the overall process. A number of factors are involved:

1. Early separation may obviate the need for additives (such as filteraid) which are costly, and costly to discard
2. Volume reduction lowers energy requirements in heating or cooling large liquor volumes
3. Pumping and stirring requirements are reduced
4. Investment is reduced since working volumes (of liquids, slurries, or solids) are much lower
5. Maintenance costs are reduced
6. Treatment or pretreatment (waste disposal) involves one mass of material with relatively constant characteristics; multiple, large-volume discard streams are avoided
7. Reagents for further treatment (adsorption, solvent extraction, ion exchange, etc.) are needed in smaller quantity
8. Some potential for reuse of a discard stream exists, since very little has been altered (no or few additives, little degradation); alternatively, a discard stream may be easily converted to an animal feed product or fertilizer

2. Cell Separation

Cell separation is performed often whether the product is intracellular or extracellular. If the product is within the cell (many enzymes, for example), the rationale is clear. If the product is in the liquid phase, cell removal is valuable because:

1. Cell mass can be treated (or sold) as an entity
2. It eliminates an interference in later processing whether or not lysis occurs
3. Potential for undesirable environmental problems is lessened (odor, contamination of other flows, aerosols)

The techniques employed in cell separation have been employed for many years:

1. Settling and flotation — in common use in waste treatment
2. Rotary vacuum filter — usually with some precoat; vacuum may or may not be required
3. Centrifugation — high speed disk machines or vertical or horizontal bowls giving a relatively moist slurry or sludge

C. Centrifugation

Centrifugation depends upon development of forces which will cause particle migration. Any particles (as cells, cell debris, coagulated protein, crystals, adsorbent, processing aids) can be separated, at least theoretically, by means of a centrifuge. In dilute systems, each particle can be treated as a single particle in an infinite field. Where more concentrated slurries are involved, particle interaction occurs and empirical relationships are often employed. In hindered settling situations, one such empirical equation is[61]

$$\frac{M_{hind}}{M_o} = \frac{1}{1 + B(VF)^{1/3}}$$

where M_{hind} = particle velocity with particle interaction, M_o = single particle velocity, VF = volume fraction of particles, and B = empirical constant.

$$B = 1 + 305(VF)^{2.84} \quad \text{for } 0.15 < VF < 0.5, \text{ irregular particles}$$

$$B = 1 + 229(VF)^{3.43} \quad \text{for } 0.2 < VF < 0.5, \text{ spherical particles}$$

$$B = 1 \text{ to } 2 \quad \text{dilute suspensions or } VF < 0.15$$

With this complexity in mind, most applications are determined by actual test runs using typical broth on a pilot machine. The test is usually run at a manufacturer's laboratory, since various machines may be evaluated in a single day. A major problem that arises in such an evaluation concerns the "typical" nature of the broth. Methods of preservation of the broth in transit, even if as innocuous as refrigeration, very often cause major changes in the particle or broth. Ease or difficulty of separation may be an artifact of the aging of the sample even if degradation were avoided. Addition of preservatives may have a major impact on surface charge and so may impact separation. While performing the centrifuge test on fresh whole broth avoids the problems mentioned, another sort of difficulty may arise. Testing 10 or even 100 ℓ of broth will require minutes or 1 or 2 hr at most. On an industrial scale, harvesting and holding a fermentor batch may cover 6 to 10 hr, even if centrifugation is the first processing step. Differences in separability have been noted in this interval. Once an acceptable separation, or separator, is selected, broths at various postharvest ages should be evaluated. Such precaution should be followed whatever the unit operation following harvest, or wherever appreciable hold time occurs.

Yeast recovery has been accomplished in that industry (including brewing) for many years. Yeast cells are most often found as single cells that are 5 μm in width (or diameter) by 10 μm in length. Bacteria are spherical or rod shaped and can range from 1 to 3 μm in diameter, or width to length (exceptions exist). Since separability is very strongly a function of particle size, it is much more difficult to separate bacteria compared to yeast. For bacteria, flow rates are reduced and incomplete separation is common. Newer machines have been constructed with forces of up to 15,000 × g. Stress on the mechanical devices is very high.

For large-scale, continuous operation, a battery of machines is used. Some overdesign is needed both to keep the plant running in case of one unit's failure and because cleanout and preventative maintenance take up a relatively high proportion of total time. The disk stack centrifuge is the design of choice.[62] Numerous disks in a stack provide a large area for sedimentation; solids move outward and collect in appropriate spaces between the stationary bowl and the rotating stack. The machines can be run so their discharge (rapidly opening and closing valve assembly at periphery) is via (1) timer, (2) signal from turbidity measurement on supernat, or (3) load or vibration detection. Manual operation is possible, but becomes burdensome for many machines. High speed machines should be equipped with a vibration detection device and special brakes which become operable whenever a predetermined excessive vibration occurs. Otherwise, the machine might self-destruct and serious injury to personnel could occur. Inertial forces that develop are very high. While solids exit at discharge nozzles, clear liquid moves to the center of the unit and leaves at a liquor outlet near the top. Separated solids must flow and so contain a good deal of the input liquid. In most applications, solid slurry is collected and washed with water. One or more reslurry steps may be set up in a countercurrent fashion. Reslurry is useful whether or not product is intracellular.

There are a number of advantages in use of centrifuges compared to other separation devices:

1.   Continuous operation possible
2.   Retention time in device short (may be seconds)
3.   Some adjustment in separation efficiency possible
4.   No additives (as filteraid) required, so both purchase cost and disposal cost avoided
5.   Smaller floor space requirement compared to similar unit operations

Recent design modifications may also be considered advantages:

1.   Closed systems, high containment levels

2. Improved metallurgy with resultant higher *g* forces and reduced corrosion of metal parts
3. Improved temperature control — obviate heat rise related to mechanical energy input
4. Capability to steam sterilize
5. Improved CIP capability

Erikson presents data on disk stack machines for enzyme purification and also shows these results for *Escherichia coli* cells and cell debris after disruption of that organism in a "typical" DNA process:[62]

	Harvest cells	Cell debris
Solids content in feed, %vol	6—7	4—5
Throughput, M ℓ/hr	3—4	1.5—2
Clarified product, %vol	0.02	0—0.02
Separated solids, %vol	70—80	40—50

E. coli cells have been separated from broth at rates of several thousand liters per hour with cell recovery of 99 + %.

Disk stack centrifuges are further subdivided into those units which have a nozzle discharge and those which discharge intermittently (with machine still running) by means of bowl or closure movement for a very short interval. Both types, intermittent or continuous pressurized discharge, have variations depending upon the specific application. General categories of separation are

- Cell broth separation — removal of cell cream or concentrate
- Cell debris removal after lysis or rupture
- Protein/enzyme precipitants or coagulants
- Solids concentration — crystal separation
- Absorbent removal
- Clarification prior to polishing filtration
- Separation of two immiscible liquid phases after liquid — liquid extraction

Decanter centrifuges should also be considered for less stringent demands. These machines have a horizontal bowl and develop lower *g* forces, hence, they are used to perform separations of more easily coagulable material or larger particles. They can be considered dewatering machines; an example is sludge dewatering in an activated sludge sewage treatment plant. Decanter units may operate in a batch or continuous mode. Horizontal rotor or bowl may be conical or conical/cylindrical. There is an internal screw conveyor that operates at a differential speed compared to the bowl; in this way concentrated solids are conveyed by the screw along the bowl wall to discharge ports. Clarified liquid passes out of adjustable ports (weirs) at the opposite end of the bowl. Variables which can be adjusted include:

Bowl speed	Feed rate
Bowl geometry/type	Feed port position
Screw conveyor pitch	Pool depth
Screw conveyor speed	Weir depth

A cross-sectional view of a horizontal decanter is given in Figure 8.

An analysis of typical bioprocesses will indicate potential use points for centrifugation. A process flow diagram is a good place to start. Immediately after fermentation of, for

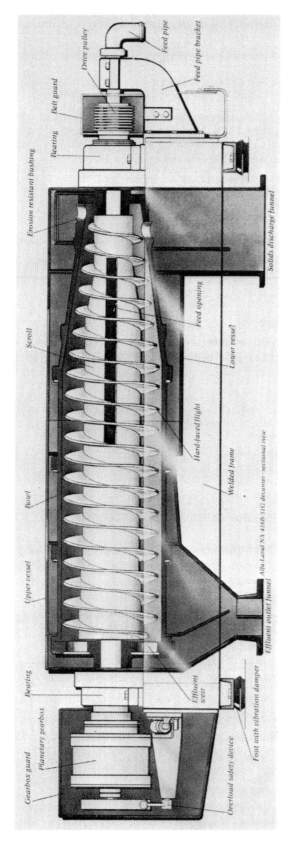

FIGURE 8.   Cross-sectional view of horizontal decanter centrifuge.   (Courtesy of Alfa-Laval.)

example, an extracellular enzyme or other product, it is useful to remove large, noncell particles. This step makes subsequent cell removal more efficient, i.e., reduced frequency of solids discharge whatever the unit operation. Very often, complex natural products are used in preparation of fermentation media; visible particles often remain in harvested broth. If centrifugation is the cell removal stage, large particles might plug nozzles or discharge orifices. Preclarification, therefore, may occur via decanter centrifuge, screening, coarse filtration, or even setting/coagulation (with or without additives). Supernatant, with most or all cellular material, will pass to a high speed disk machine for cell and smaller particle separation. The supernatant from this step will contain some low level of cells and cell debris. A polishing filtration step may or may not be needed, depending upon subsequent steps. It should be obvious that if an intracellular product were desired, the cell paste or solids from the disk centrifuge would receive further processing. For the extracellular product case, the clarified stream will be processed. Assume some form of concentration is used (membrane system, evaporation, salting out, selective crystallization) and some solids formation occurs. Centrifugal separation is possible once again. If a wet slurry results at some subsequent step, dewatering can be accomplished readily prior to product drying.

In large-scale production of amino acids via fermentation, centrifuges are, or have been, used for:

1. Cell removal (and by-product sale or use)
2. Cell/amino acid slurry separation with intermediate reslurry wash and separation (isoelectric precipitation)
3. Precipitated inorganic salt removal (and by-product use)
4. Fractional centrifugation of salts of certain amino acids
5. Removal of adsorbent or carbon particles (purification or decolorizing steps)
6. Amino acid crystal dewatering (prior to drying)
7. Desludging/solids concentration in waste treatment

Problems with use of centrifugation are

1. High capital investment plus high maintenance cost
2. Heat generation at high speed may be a problem for certain biological products
3. Aerosol generation common — removal or filtration of particles may be needed
4. High noise levels common — special protective screens or cabinets may be required
5. Rebalancing of rotating elements most often must be performed at manufacturer's shop — costly and time consuming

One of the more promising methods of reducing downtime and repair costs is use of vibration analysis, sometimes referred to as signature analysis. Portable apparatus can be used to measure the distinctive vibrations generated by moving elements. Every moving part of operating equipment generates a characteristic frequency which is used to determine whether undesirable changes have occurred with time. Most importantly, the equipment remains in use during testing and no downtime is incurred during inspection. Planned shutdowns can be taken; very often an analysis of the readings will direct maintenance personnel to the problem area and parts and equipment will be in place to facilitate repair. Other usual precautions should be taken to ensure stable operation:

- Stable feed (composition, rate)
- Uniform torque
- Lubrication checks
- Noise level checks
- Abrasion checks
- Corrosion or erosion (especially trunnions and bowl)

One manufacturer has presented approximate cost figures (mid-1985) with the understanding that special designs and special requirements are common and that figures shown can only be taken as a guide.[63]

Solids Ejecting Centrifuges

Alfa-Laval model	Capacity estimate (ℓ/hr)			Price ($000)
	E. coli cells	Cell debris	Protein	
BTAX-215	3500	1750	5000	165
BRPX-617	1750	875	2500	175
BRPX-413	1050	525	1500	140
BRPX-309	350	175	500	115
BRPX-207	210	105	300	80
BTPX-205	175	87	250	55

Pricing is based upon:

1. Basic machine, motor, tools
2. NEMA 4 motor starter and machine control
3. Simple process piping
   Feed and back pressure valves
   Sight glasses
   Flexible hoses
   Pressure gauges
4. Tanks, external pumps, turbidity meter not included

BTAX and BTPX are designed for biotechnology applications (special materials of construction, special seals, capability for steam sterilization). BRPX are general purpose machines.

Nozzle Centrifuges — Yeast Applications

Alfa-Laval model	Capacity (m³/hr) bakers' yeast	Price range ($000)
DX-409	25—45	40—45
FEUX-510	40—70	70—80
FEUX-512	50—80	90—100
FEUX-420	120—200	290—300

Decanter Centrifuges

Alfa-Laval model	Diameter (mm)	Length (mm)	Relative capacity	Price range ($000)
NX-309	230	700	1	60—70
NX-414	353	860	1.5—2	80—90
NX-418	353	1460	3—5	110—120
NX-429-32	620	1850	10—15	240—250

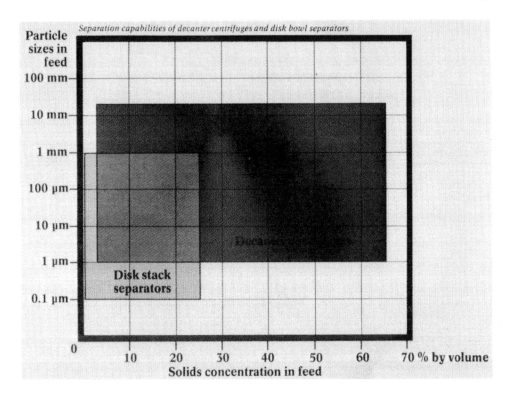

FIGURE 9. Application ranges of disk stack and decanter centrifuges. (Courtesy of Alfa-Laval.)

It is possible to present generalized particle size and solids content data while indicating regions of suggested utility for disk stack or decanter machines. Such a graphical presentation is given in Figure 9. For small particles and solids content of 25% by volume (or less), a disk stack separator is suggested. For larger particles (greater than 1 mm) or elevated feed solids, a decanter centrifuge is suggested. A large region of overlap exists.

D. Cell Disintegration

Should the product remain as an intracellular entity, some method of cell rupture or disintegration is needed so that the product is released into the suspending medium. An ideal solution would be carefully controlled autolysis of the producing organism, once peak product concentration has been reached. A prophage could be present, or inserted, which is triggered by some externally controlled variable such as pH or temperature. This is a highly idealized situation. A more likely microbiological route would be controlled phage infection either pre- or postharvest. Obvious requirements would be ease of handling, no cross-contamination of other fermentations, rapid disappearance of viability or infectivity in air, and no effect whatever upon the product. If aerobic conditions were required after infection to ensure phage replication, one can be certain that a very high phage count will exist in ambient air unless specific precautions are taken.

While the microbial route is attractive from energy and other cost considerations, the deficiencies noted are major disincentives. Mechanical devices are often used even though investment and operating costs are high. High fluid shear forces must be achieved to cause cell rupture. Bead mills may be employed, but homogenizers are more commonly used. In a typical homogenizer, a very high upstream pressure is achieved by means of an internal pump and a large pressure drop occurs across a small nozzle or orifice. Throughput is not normally very high.

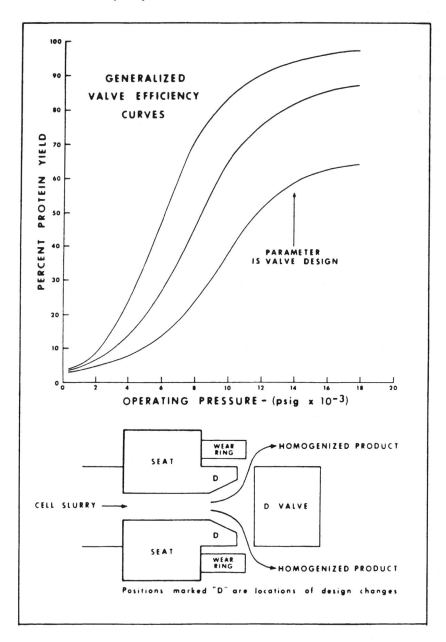

FIGURE 10.   Generalized homogenizer valve design and percent protein yield vs. operating pressure
as a function of valve design.

The goal is a single pass operation that maximizes protein or enzyme yield as a percentage
of the total cell mass while minimizing power consumption. One manufacturer (APV Gaulin,
Inc.) has reported on recent work with various valve designs at operating pressures up to
20,000 psig. Both valve geometry and homogenizing pressure influence overall efficiency.
(See Figure 10, which shows areas of design modification in homogenizer valves and impact
of design and pressure on protein yield.) The measure of efficiency is available protein
release using a slurry of bakers' yeast; at pressures near 10,000 psig with newer valve
designs, yield has been increased from 45 to 75%. Further, increases in pressure reduce the
need for multiple passes. Recycle of partly processed cell mass is undesirable, since through-

put is reduced, temperature effects are normally detrimental to the product, mechanical wear is a problem, and cell debris is also reduced in size rendering later separation more difficult. Manufacturers' information is available for various microorganisms and studies have been performed on mechanisms of disruption and efficiency of various techniques.[64] Enzyme release from yeast and studies of a number of key variables have been made.[65]

A generalized estimate of cost can be made. One can assume that cell mass in the feed slurry is 15 to 25%; however, actual experimentation on a pilot scale would be needed to confirm sizing.

GPM of feed	Horsepower requirement	Equipment cost (Mid-1985, uninstalled)
5	30	$35,000
10	60	50,000
20	125	70,000

A third method of cell disintegration involves use of an added chemical. The chemical may be a disinfectant (quarternary ammonium compound, chlorine, or its derivatives), an antibiotic that interferes with cell growth or reproduction, an enzyme that attacks cell walls or membranes, or a surfactant that causes lysis. Once again, the chemical or enzyme of choice should have no effect on the product and should leave cell debris in a state that can be readily separated.

E. Filtration

Filtration is not only utilized in product purification, but is a critical unit operation in air sterilization. This aspect is discussed elsewhere, but a few general rules apply in any filtration application. The first concerns the quantity and minimum size of the particles to be removed from any flowing stream, whether gas or liquid. The finer the filter, the greater the cost (in most applications). Other operating criteria which should be considered early in design are

Viscosity	Cake deformability	Temperature
pH	Cleaning requirements	Pressure drop
Corrosion rate	Flow rate	Filter medium

Sizing depends upon process conditions, throughput rate, and desired process result.

General reviews of filtration theory and practice are available.[66,67] The first reference details small-scale tests needed prior to equipment selection.

In fermentation practice, it is often necessary to add a filteraid (material containing noncompressible particles). The ideal filteraid would have these characteristics:

Low bulk density
Low cost
Porous
Incompressible
Low cost disposal
Chemically inert
Capable of simple regeneration
Nonblinding
Constant size

Examples of common filter aids are diatomaceous earth, carbon, pearlite, and clays.

Commonly found filters in fermentation plants include:

Conventional rotary drum
  Supported cylindrical drum
  Cake on outside of rotating drum, removed by scraping
  Constant level control
  Cake washing
String discharge filter
  Modification of above where an endless string lifts off cake to discharge it
  Cake must be sufficiently stable to be removed by strings
Continuous vacuum precoat filter
  Adaptation of no. 1 above suitable for gummy or colloidal suspensions
  Layer of filter aid formed prior to start of run
  Advancing knife edge removes thin film of deposited solids plus thin layer of filteraid
Precoat pressure filter
  Multiple leaves, plates, tubes with precoat added prior to run
  May be readily automated (automatic cake discharge and reforming)
Disk filter, cartridge filter
  Disks of paper or plastic compressed on multiple screens or plates
  Extreme clarification (including sterilization) is possible
  High surface area for space occupied
  Large potential range for particle size removal, but usually limited to low viscosity, low
    solids streams

A review of filter design and practice is given by Boss.[68] A useful guide to filter selection is given. The continuous vacuum precoat filter is a general-purpose unit that offers flexibility without major equipment redesign. Liquid is removed through the filter face, through a vacuum line, and into a receiving vessel or drum. Drum internals can be configured and altered easily so that vacuum is applied only to fixed positions of the drum face. Solids must be removed with each rotation, or fresh surface (with or without filteraid) exposed, so that a clean and unplugged surface is available for renewed filtration. As drum rotation occurs, solids are sloughed or cut off and are conveyed (slurry or mechanically) to a solids handling system. Variables that can be changed with minimal delay or cost include: filter cloth, filteraid, vacuum, drum speed, rate of cake removal, and type and rate of cake wash. These units have been used to separate cells from fermentation broth. Mycelia and yeast are relatively large cells and filter separation is both simple and advantageous. Bacterial or coagulated protein separations are more difficult and require addition of filteraid.

Filtrate enters a drainage pipe and passes into a vapor-liquid separator. The receiver is a sealed tank; vapor is removed via a vacuum source. Cake discharge depends upon cake characteristics. Simple, easily separable, and crumbly cakes can be removed with a doctor blade with or without air blowback. A string discharge (parallel, closely placed string moves around filter drum and onto two or more rollers before return to drum) is useful for thicker, somewhat more cohesive cakes. A continuous belt discharge is used for more complex cakes which tend to adhere to, or blind, the filter cloth. The belt is usually cleaned (by washing) after cake is discharged and before the belt returns to the drum. When a precoat is used, a moving knife edge removes a fraction of the cake. In this way, fresh bed surface is exposed to liquid slurry to be filtered.

The continuous rotary vacuum filter is available in many materials of construction and in sizes from 3 to 12 ft in diameter, with face lengths from 1 to 24 ft. Special designs and sizes can be ordered. Filtration rates vary from less than 10 GPH/ft^2 to highs of 100 to 150 GPH/ft^2. Moisture content of cake may be as low as 25%, up to a high of 75%.

Cost data (purchased cost vs. filter area or batch size) are given in Reference 67 (data as of January 1979) and cost data are also given for many filter types in a log-log plot of cost vs. total filter area in Reference 66 (Figures 19 to 116, data as of early 1982).

F. Intermediate Purification

The entire process of purification of a fermentation product may be summarized as water removal and impurity separation. As such, both the sequence of events and unit operations used should be selected on the basis of how much separation can be done simply (read, economically) before more complex (read, costly) separation steps are employed on a smaller mass or volume of concentrated product.

After the clarified broth is prepared, many intermediate concentration steps are available for water removal or selective separation. These include:

Evaporation	Ultrafiltration
Distillation	Adsorption
Precipitation	Ion exchange
Solvent extraction	

The criteria of choice at this point center upon available installed equipment, new capital and operating costs, and the need for subsequent processing steps (how many and how costly). It is likely that alternatives exist; selection will depend upon company experience and expertise with similar processes and efficiency or effectiveness of the possible methods. One aspect of the selection is often neglected, but ongoing regulatory pressures are forcing the earlier consideration of environmental impact and resulting effluents. In fact, it is not too far-fetched to present a scenario where the most process-efficient (yield/quality) concentration step is discarded, because personal protection requirement is too stringent or because a resulting waste stream must be transported too far away for disposal. Emissions of certain solvents may not be permitted and containment would be onerous. Environmental impact becomes one of the critical selection criteria.

G. Evaporation

Evaporators remove a solvent vapor from a liquid mixture or solution by means of some sort of energy input to the process. A change of state occurs with vapor transfer, and in almost every instance condensation takes place at a subsequent step. Condensed liquid may or may not be recovered. The vast majority of applications of evaporation in the biotechnology industries involves water as the solvent. Furthermore, steam is normally the heating medium. Therefore, the unit operation of evaporation involves heat exchange generally across a metallic interface. Considerations of energy efficiency and conservation are of major importance. Since water removal is a key requirement (broth concentration may be as high as 15% by weight of desired component, but usually titer is much lower), and since retrofitting is time consuming and costly, initial plans should include a highly efficient evaporation system with provision for facile expansion. Modern systems involve use of multiple effects. The vapor from one effect (with or without treatment) is used as heating medium for a subsequent effect where liquid boils at a lower temperature and pressure. The number of effects in a specific design depends upon allowable temperature range, cost of energy, capital cost increments, and desired concentration.

Most evaporators consist of three parts: a heating section (calandria), apparatus for liquid-vapor separation, and a structural assembly to contain these elements and separate process and heating fluids. Heat transfer surfaces may be internal or external, but whatever the design, fluid must be circulated across or through the surfaces to affect heat exchange. Many designs are available.

One type, the falling film, is suitable for fermentation broth applications. Size can be very large, liquid holdup is low, multistaging is simple, excellent heat transfer rates are achieved over a wide range of conditions, and plant area is minimal. Some level of viscosity and moderate fouling characteristics can be tolerated, since the design depends upon a thin fluid layer falling by gravity from top of tubes to the bottom (concentrate). Liquid distribution is important, since bare or hot spots are undesirable and are to be avoided. The uniform temperatures and short residence times mean that heat-labile products can be concentrated. The best applications of long-tube falling film, vertical evaporators are for clear liquids, high evaporation loads, corrosive solutions, and low temperature operation with low temperature differences. Disadvantages relate to sensitivity (ease of upset) to changes in operating conditions and necessity for proper feed distribution. Advantages and disadvantages for many evaporator styles are given by Rubin et al.[69]

Plate-type exchangers and spiral plate exchangers are often used for media sterilization or poststerilization heat recovery (and cooling of the sterile media). Advantages are highly turbulent flow, self-cleaning of heat transfer surfaces, various materials of construction, capability of handling some slurries, and rather low space requirement. Plate and frame units can handle heat-sensitive, viscous, foamy, solids containing fluids. Difficulties do occur in gasket selection, gasket assembly, and equipment reassembly. Gaskets are essential to prevent liquid interchange and/or exposure of liquid or steam to air. It is not a simple matter to regasket such a unit or to reassemble it after an inspection. Plate heat exchangers also offer the advantage of relatively simple addition or subtraction of transfer area.

Thin-film or scraped surface evaporators, while costly, are found in many biological purification plants. Thermal degradation is minimized, since residence time can be measured in seconds. In this unit, liquid enters above the heated zone and is moved by mechanically driven blades that effectively spread the fluid across the interior, thermal surface of a cylinder. Scraper blades may be spring loaded. Hallmarks of the design are high surface renewal, very high liquid turbulence, and very low liquid holdup. A carefully designed drive, seal, and rotor assembly are essential and tolerances are very low, hence, the high capital cost. Since viscous materials as well as solids-containing liquids can be handled, and since fouling tendencies of the fluid feed have little impact upon concentration, special and problem mixtures can be processed. Generally, a thin- or wiped-film evaporator is not the first choice for this unit operation. Since viscous or heat-sensitive materials (often found in biological processing) can be concentrated under atmospheric pressure or under high vacuum, there is a need for such a design.

A review by Mehra[70] discusses all types of evaporators, gives design details, and reviews pros and cons for the various styles. A more detailed economic analysis of four different concentration processes, at three different water removal rates, has been published.[71] The four processes are triple-effect evaporation, mechanical vapor-recompression evaporation (MVR), reverse osmosis, and freeze concentration. While the latter is a drying method, also, in this report input solids were assumed at 5% and output solids at 30%. Water removal rates were 450, 4500, and 45,000 kg/hr. A careful and detailed listing of assumptions is given (costing is 1980/81 averages) and a tabulated breakdown is given for fixed capital and operating cost. Further, many references (1971 to 1981) are given for papers that have published capital cost and operating cost estimates for liquid food products (juices, whey, coffee). Costs are updated to 1981 basis. Table 18 rounds figures and summarizes results for the processes studied.

H. Membrane Processes

In the last two decades, great strides have been made in applying membrane technology to commercial biotechnology. Use of membranes offers cost-effective and energy-efficient

Table 18
CONCENTRATION ECONOMICS (FOUR PROCESSES COMPARED)

Water removal (kg/hr)	450		4,500		45,000	
Process	Capital	Operating	Capital	Operating	Capital	Operating
Triple effect evaporator	0.32	9.4	0.68	2.0	2.27	1.0
MVR	0.31	8.7	1.02	1.7	3.65	0.5
Reverse osmosis	0.43	11.5	1.88	3.7	11.55	1.8
Freeze concentration	1.72	24.8	4.70	6.0	30.90	3.7

Note: Capital cost is in millions of dollars; operating costs are in cents per kilogram of water removed.

methods of concentration and separation. Four main areas have been identified where membranes have high potential:[72]

- Sterilization of bioreactor feed streams
- Downstream processing of spent medium
- Immobilization of biocatalysts
- On-line monitoring of fermentation (membrane electrodes)

It is possible to add pyrogen removal from water used in final product preparation and packaging. There are a number of different applications which relate to membrane structure and porosity, driving forces for mass transfer, and composition of streams. Certain of these processes are

- Microfiltration
- Ultrafiltration (UF)
- Reverse osmosis (RO)
- Membrane distillation
- Pervaporation
- Electrodialysis

Physical arrangements may include tubular membrane modules, capillary membrane modules (packaged within a single retention chamber), plate and frame systems, hollow fiber modules, and spiral wound membrane systems.

The major application of membranes has been in the area of downstream processing. The first separation operation on harvested whole broth is normally filtration or centrifugation, or a combination of the two. Whether or not cell lysis occurs first, an intial step of cross-flow microfiltration can be used to either remove cell debris or fragments or to concentrate whole cells. Energy requirements are low, additives are not required, and temperature effects are minimized. Furthermore, maintenance and operating costs are reduced if an appropriate membrane material is selected.

Microporous membranes can be used to separate components by size. Clear supernatants and high purity streams can be obtained readily. Cross- or tangential-flow filtration (fluid flow is at right angle to direction of flow) minimizes cake formation and plugging.[73] Microporous membrane hollow fiber cartridges have been designed specifically for cell harvesting. The cartridges are sanitary in design, heat sterilizable, and bubble-point testable. Essentially complete recovery of microorganisms is possible. In any membrane process using fermentation broth, presence of processing aids must be considered. Most broths contain defoamers used in the course of the fermentation. These surfactants could have a serious detrimental effect on the membrane; this is especially true if broth goes directly to a membrane

Table 19
COMPARISON OF MEMBRANE PROCESSES REVERSE OSMOSIS AND ULTRAFILTRATION

	Reverse osmosis	Ultrafiltration
Pressure	High	Low to moderate
	(500—1000 psi)	(up to 300 psi)
pH	5—7	2—12
	(3—11 possible)	
Temperature	Up to 100°F	Up to 150°F
	(to 170°F possible)	(to 170°F possible)
Material	Cellulose acetate	Polysulfone
Pore size cutoff	MW of 100	MW of 10,000
(usual)	(to 500 possible)	(higher cutoffs, also)
General process	Concentration	Fractionation

system for its initial processing step. Special cleaning techniques are available. An alternate technique is to test and select defoamers for best overall system response, i.e., selection based on best operating economics overall, not just for prevention of foam loss during fermentation.

Subsequent steps in conventional technology might include evaporation, distillation, and/ or absorption. One or all could be replaced by a membrane apparatus. Volatile solvents (or products of fermentation) could be removed by pervaporation. Low molecular weight salts can be separated by one or another membrane process. Electrodialysis might replace ion exchange. Water can be removed from a process stream by UF or RO. Since there are many cases where no chemical additive is employed in a membrane operation, a potential for recycle arises. Unused substrate can be recycled to the fermentor, with or without additional processing. Here, the key determinant is the cost of medium constituents; expensive precursors or cofactors that are recoverable would be recycled. Whole cells can also be recycled to increase reactor productivity. Since sterility can be maintained in membrane separation, there is an opportunity to reduce cycle time or, at least, to save on substrate to cell conversion. Final product purification, in a conventional flow diagram, might include crystallization, selective adsorption, or some type of chromatography. Membranes can be selected which will concentrate on desired product while rejecting impurities. This may involve sequential steps with different membranes.

Membrane reactors or membrane separators as an integral part of a fermentation system (recirculation of whole broth) can be operated. These systems have gained increasing interest as mammalial cell culture and production of monoclonal antibodies have become commercially viable. Once again, membrane systems offer the positive attributes of no shear effects, no temperature or pH effects, and potentially high selectivity.

Although there is overlap in use and novel systems are being developed with rapidity, Table 19 lists some characteristics of RO and UF.

Selected cost analyses and comparisons of UF and RO in fermentation processing have been presented.[74] Combined systems or easily convertible systems have been used for broth preconcentration and also for recovery or recirculation of spent broths. The advantages claimed for a new series of RO membranes are

1. Concentration up to 30% solids at ambient or higher temperatures
2. Highly energy-efficient water removal with 100 to 200 kg water removed per 1 kW hr
3. Simple operation: one pressure or flow control valve controls plant operation
4. Modular construction, stainless steel modules, easy to clean, assemble, install, and amenable to expansion

Table 20
CAPITAL AND OPERATING COST OF MEMBRANE PROCESSES

Reverse osmosis (RO)		**Ultrafiltration (UF)**
5285 GPH	Plant capacity	15,850 gal batch/30 hr
92	Power consumption (kW)	32
2× (5—10% TS)	Concentration	8—10×
392	Capital cost ($000)	122
5000	Operation (hr/year)	4500
$34.4 M/year	Membrane replacement	$3.5 M/year
27	Annual power cost ($000)	7.8
$6000	Chemical cleaning cost (annual)	$800
$0.511/100 gal water removed	Operating cost	$0.511/100 gal feed

Note: One pound sterling = $1.35 and power cost = $0.054/kW hr.

Design characteristics and costs are given for an RO plant that concentrates spent fermentation broth (yeast production). Data for a UF plant separating and concentrating an enzyme solution are also given in the reference noted. Table 20 summarizes the data with units converted as shown. It was noted that operating cost for RO concentration is less than half that calculated for four-stage evaporation to achieve the same result (assuming 5 lb water removed per pound of steam).

A review article (22 references) discusses specifications of commercially available ultrafiltration membranes and modules.[75] Some 11 manufacturers are covered and cutoffs and retention for available types are listed. Some theoretical background is given and there are clear cutaway views of various geometries as well as pictures of three industrial installations. Applications discussed are

- Dairy industry (whey fractionation, cheese milk concentration)
- Electrocoat paint recovery
- Pharmaceutical applications
- Ultrapure water production
- Medical uses (artifical kidneys, plasma protein separation, artificial organs)
- Effluent treatment (including textile size recovery)

Some size and capacity data are given, but no capital or operating cost information is given.

Another review article centers on scale-up considerations.[76] The importance of maintaining several key variables constant during scale-up is emphasized. The critical parameters include:

1. Inlet and outlet pressures
2. Recirculation velocity
3. Product content and concentrations
4. Membrane type and membrane configuration
5. Overall process time

Certain details must be monitored carefully to be sure these results are obtained:

1. Hollow fiber diameter should be constant.
2. Data from lab/pilot cassette systems cannot be used directly to size spiral elements.
3. Test samples must be representative of manufacturing runs (note that this point is emphasized for almost every unit operation in downstream processing).
4. Test run time should be monitored (membrane fouling effects).
5. Full-scale equipment must have properly sized pumps and manifolds.

In this analysis, bioproducts which are processed by membranes are separated into two groups:

1.   Dissolved macromolecules, as enzymes, plasma proteins, viruses, hormones, peptides, lymphokines, interferons
2.   Cellular products, as antibiotics, enzymes, toxins, alcohols

Case studies are presented for albumin concentration/diafiltration, interleukin-2 concentration, cell debris removal for growth hormone isolation, and cell harvesting/cell washing (*E. coli*). Cross-flow filtration is compared (process result and economics) to centrifugation on a laboratory-scale cell separation procedure. Hollow fiber filtration is superior on both counts. A more detailed economic comparison is made for production scale harvesting of *E. coli*. A simple medium is assumed with broth containing 40 g/ℓ feed to separation process and product containing 200 g/ℓ. An Alfa-Laval continuous disk-type machine is compared to a hollow fiber system based on Amicon 0.1-μm microporous membranes. With the assumptions made, the hollow fiber system would cost $45,000 initially with an operating cost of $2.27/k$\ell$, while the centrifuge system would cost $150,000 initially with an operating cost of $2.99/k$\ell$. The difference in depreciation is, however, $0.87/k$\ell$. Membrane cost is taken as $157/m^2 with a life of 1 year. A generalized bioproduct flow diagram is given which shows that membrane processing can be employed at every step of the operation, beginning with medium purification (fermentor feed) and ending with pyrogen removal for water used to wash vials or dissolve finished product.

An article on DDS (De Danske Sukkerfabrikker) systems for UF and microfiltration (MF) for downstream processing is also available.[77] All UF and MF membranes made by DDS and recommended operating ranges are given. A process for separation of proteins (MW of 2000 to 4000) after enzymatic blood cell hydrolysis is given. Once again, an economic comparison is made between centrifugation and UF. Assumptions include operation for 200 days/year, electricity at $0.059/kW hr, 25 k$\ell$/8-hr day. Protein yield is rather different. For centrifugation, yield is 69%, with water addition at 74% of 25 kℓ/day. For UF, yield is 80%, with water addition at 40% of 25 kℓ/day. Investment for centrifugation is given as $353,000, for UF as $141,000 (assume $1 = 8.5 D.kroner). Membrane life is taken as 1 year with cost of $118/m^2. Operating cost for centrifugation is $9765/year, with only electricity, general maintenance, and manpower included (depreciation excluded in both cases). For UF, running cost (electricity, membranes, manpower) is $19,880/year. However, the value of the additional protein recovered is $77,000/year (38,500 kg valued at $2.00/kg). Other commercial possibilities include UF concentration of xanthan containing whole broth (2% w/w xanthan gum plus organisms to 9%) and wine clarification procedures are described.

Food proteins have been recovered by UF for many years. The laboratory and pilot plant studies needed to properly evaluate centrifugation and UF are presented in a paper by Hoare and Dunnill.[78] The isoelectric precipitation of a protein (soy protein, in this case) can only be affected in an efficient manner if the following studies are made:

1.   Extent of precipitation and denaturation as a function of acid type and pH
2.   Extent of shear damage to proteins
3.   Sensitivity of precipitated protein to mechanical forces during formation and separation steps
4.   Sensitivity of protein to gas-liquid interfaces
5.   If mechanical forces are disruptive, what type of pump/recirculation device would be optimal
6.   Effects of aging (mixing, temperature, acoustic conditioning) on particles

Physical measurements and laboratory kinetic studies can be used to predict behavior in industrial processing.

Generalized cost information has been discussed. While each installation will require a specific estimate, some ranges for investment and operating cost are useful. From discussions with manufacturers, a system with UF area of 10,000 ft² (or 929 m²) and having a permeation rate of 12 gal/ft²/day (or 489 ℓ/m²/day) would cost from $600,000 to $1 million to install (prices mid-1985). Conventional membranes are assumed in use. Operating cost range is from $6 to 10/1000 gal of permeate. Membrane replacement costs and cleaning chemical costs are included in the estimate.

A recent book covers many areas of membrane separations.[79] Cell harvesting is reviewed and specific applications to blood and plasma fractionation are given. Chapters on applications in plant tissue culture, food technology, and artificial organs are included.

I. Final Purification

The purification of the new generation of biodynamic compounds will, in general, involve equipment which can be defined as "large laboratory scale". An industrial-scale centrifuge or dryer for large volume, relatively stable fermentation products is very different from the laboratory model initially used to affect the separation. On the other hand, highly complex and often unstable compounds or mixtures require equipment which may not be commercially available. An adsorption on a selective column can be scaled to commercial size, but both volumes treated and cost of the support or selective carrier mitigate against very large scale-up factors. A disaster is unpleasant in an 8-in. × 6-ft column; it is totally unacceptable in a 4 × 30 ft column.

Pharmaceutical product cost (newer generation) is such that very expensive devices or complexing agents may be used. Regeneration is often unnecessary from an economic viewpoint. Since quantities are usually small and activity high, yield becomes a major determinant. Four separate purification steps, each at 80% yield, means that about 60% of the starting material has been lost by the time acceptable product is made. The key design parameter becomes selectivity. Separation must be achieved in one or two steps with elution or removal of most of the desired material, by itself, in physiologically active form. If human use is contemplated, pyrogens must be absent. Therefore, high affinity substrates should be sought. A good deal of success has been obtained in the creation of selective, high affinity adsorbents (use of monoclonal antibodies as an example) for this application.

A certain number of novel technologies may be applied if (1) small volumes are to be processed or (2) a very costly final product results. Among these newer processes, one can list

- Liquified gas extraction or supercritical fluids
- Electrokinetic transport
- High gradient magnetic separation
- Specifically bound adsorbents (examples are immobilized cells or enzymes)

Chromatographic methods include

- Size exclusion chromatography (gel filtration)
- Ion exchange chromatography
- Affinity chromatography
- Adsorption chromatography

One company manufactures a modular system for any of the above separations, with bed volumes of a few hundred milliliters to volumes of over 50 ℓ. Diameters are 13 and 252 mm and column lengths may be 15, 30, or 60 cm. Operating pressure range is 0 to 3 bar and biological compatability (nonreactivity) is claimed.

A number of recently developed techniques are in use for specific purifications with high resolution. Specific and reversible interactions between biologically complementary molecules are used to selectively separate a desired bioactive compound. A number of these techniques are reviewed by Lowe.[29] Some 59 references are given and most are quite recent. The techniques discussed are

- Affinity chromatography — a complementary ligand attached (with or without a spacer molecule) to an insoluble and inert matrix. The adsorbent acts as a chromatographic column with extreme selectivity. Plasma proteins and enzymes have been so purified.
- High performance liquid affinity chromatography (HPLAC) — a combination method employing speed and resolving power of HPLC with biological specificity of affinity chromatography. Enzymes, albumin, antigens, antibodies, glycoproteins, nucleosides, and nucleotides have been purified by this technique.
- Affinity partition — biospecific interactions of proteins with complementary substrates in aqueous two-phase systems composed of polyethyleneglycol and dextran or salts. Method has been used to purify specific cell membranes, albumin, and various enzymes.
- Ultrafiltration affinity partition — the affinity adsorbent is mixed with the crude material to be purified and the mixture is circulated within a hollow fiber or membrane unit having a mol wt cutoff of 1 million. Effluent (permeate) will contain impurities. Enzymes have been purified using this method.
- Affinity precipitation — bifunctional ligands (as nucleotides connected by a spacer) can precipitate enzymes by binding to active sites. Morphologic and topographic studies of enzymes are also possible. Enzyme purification has been achieved.

Cost information on large-scale chromatography is scarce. However, all such information is not absent from the open literature. Curling and Cooney[80] discuss process chromatography on a commercial scale. Details on control and operation of the system are given; these are very useful for anyone contemplating scale-up of chromatographic separations. The general headings covered are

1. Requirements for process water and eluent buffers (freedom from pyrogens and bacterial contamination)
2. Maintenance of cleanliness or sterility
3. Maintenance and regeneration of gel media and their integrity
4. Control parameters
5. Products of chromatographic systems and formulation into pharmaceutical products
6. Requirements for operating areas and for personnel

A case study for albumin purification from human plasma is presented. The albumin is produced with purity of over 99% and overall yield is approximately 90%. An interesting point is made concerning column maintenance. At the final stage of purification, albumin is chromatographed on Sephacryl S-200. Maintenance of this gel is achieved by a simple wash with 0.1 M NaOH once a week. For long-term storage, 0.1 mM NaOH is used. Advantages of this reagent are many:

1. Presence in final product presents no problems.
2. It is effective as a disinfectant.
3. It acts to solubilize proteins and lipids and so will remove them from the gel.
4. It is inexpensive and readily available.
5. Disposal is simple.
6. It is compatible with media used.

Other bacteriocidal agents do not have all these advantages. The overall cost analysis shows that for each 1000 ℓ of plasma fractionated, gel consumption is

0.8 kg Sephadex G-25 coarse	($0.25)
10 ℓ DEAE-Sepharose CL-6B	($1.65)
1.2 ℓ CM-Sepharose CL-6B	($0.20)
1.6 ℓ of Sephacryl S-200 superfine	($0.20)

If each unit of albumin product is taken as 100 mℓ of a 25% w/v solution, the total cost for gel media is $2.30/unit. The breakdown for each gel is given above in parentheses. The overall cost breakdown is given as (per unit):

	Dollars
Chemicals	0.35
Filters	1.55
Membranes	0.15
Electricity	0.02
Process water	0.05
Separation media	2.30
Total	$4.42

Manpower and maintenance requirements should be low (but are not included above). Neither is depreciation, cost of plasma, or quality assurance.

Dwyer[81] discusses HPLC and scale-up of that procedure. He summarizes a cost tendency in biotechnology. For "older" antibiotics, the cost ratio between fermentation and recovery was approximately 60:40. For third- and fourth-generation antibiotics, the ratio is about 40:60. For DNA products, the ratio may be 20:80 or even 10:90. The paper shows a figure prepared by A. D. Little that compares concentration of bioactive or fermentation-derived product in starting material (g/ℓ) with selling price of that finished product ($/kg). The relationship is remarkably linear. Ethanol, citric acid, and MSG at about 100 g/ℓ sell for from $0.40 to $1.00 + /kg. At the other end of the plot, factor VIII, urokinase, and other therapeutic enzymes are present at near 10^{-6} g/ℓ in starting material and purified product sells at 10^8 to 10^9/kg. Process HPLC is then discussed as a preferred unit operation to affect economic separation in kilogram preps. A key point is that large-scale HPLC equipment is not merely a larger version of an analytical device. Production scale equipment is operated to achieve high throughput rates and give product at a reasonable or acceptable purity. Dwyer states that operating costs for process HPLC fall in the range of $30 to $200/kg with mobile phase consumption a major cost factor. The author's company (Millipore Corp.) has established a new systems division to combine in-house membrane technology with HPLC capability to speed isolation and purification of new biologicals. An albumin purification system is described that can process 5000 to 6000 ℓ per shift; microprocessor control of all functions is claimed.[82]

A pilot-scale HPLC that is mounted on a movable skid (2.5 × 3 × 6 ft high) is available from Millipore. Column cartridges are changeable and different products can be processed; cartridges can be stored for reuse. Rates of several kilograms per hour can be reached and excellent resolution can be achieved. The KILOPREP™ 1000 chromatography system can operate at pressures up to 1000 psi; the system is explosion proof and the unit is designed for an industrial environment. Minimum operator training is required. The basic unit consists of a chromatography module (8 × 24-in cartridge) including a solvent pumping system and a refractive index detector. Cost is $87,500 (early 1986) and increases to $92,500 with a UV detector. Cartridges are not included. Cartridges are interchangeable and removable; price range is $2500 to $10,000 based upon packing material selected.

Other preparative scale systems are available. Varex Corporation has a basic system (PSLC-100) with these capabilities:

- Microcomputer control
- Manual/auto operation
- Binary/ternary gradient
- Isocratic mixing
- Precolumn capability
- Sample recycle
- Solvent recirculation
- Five-port fraction collector
- Automatic sample introduction
- Safety alarms, automatic shutdown

The basic system is $28,900 (late 1985) with additional costs in the range of:

Cost	($000)
Pumps (dual- or triple-head diaphragm)	7—10
Columns	
(Axial compression system)	3
(Axial compression column)	5—10
(Empty SS columns)	0.4—3
Recorders	0.7—1
Fraction collectors	3—4
Detectors	3—7

A complete system, therefore, would cost somewhat in the range of $40,000 to $60,000. In late 1985, a fully automated system with UV detection and a 4 × 30-cm axial compression column plus peak slope detection would cost $56,400 (PSLC-100R).[83]

Scale-up of chromatographic separation, with some cost data, is discussed by Skea.[84] A simulated process system was set up to emphasize studies of column life, column rejuvenation procedures, and economics of the overall separation. Cephalosporin C in fermentation broth was the crude feed. Broth pretreatment consisted of cell removal via microporous tangential flow filtration. After loading data determination and column life and rejuvenation experiments were completed, larger-scale process economics could be determined. One "Kiloprep" cartridge (Millipore) is to be used with loading of 450 g/cycle. Recovery is 98.5% and mobile phase flow is 3 ℓ/min. With various operating assumptions made (based upon small-scale experiments), the system production rate is 1555 kg/year. Labor, operating costs, and depreciation are summed to show that it will cost $169.80/kg to produce a product that has a purified value of $550/kg. After adding value of the crude ($1221.50/kg), the "value" of production becomes $391.30/kg. Contribution of the large-scale chromatography is $158.70/kg ($550 less $391.30) or the total contribution is approximately $247,000/year. Large-scale HPLC may not be the purification method of choice; however, the processing value is quantitatively established and an economic comparison to alternatives (which must be costed accurately) can be made.

A summary of preparative and process chromatography equipment with estimated (1986) costs is given below:

Company		Cost ($000)
Beckman	Prep-350 gradient chromatograph, reverse phase, fast affinity, hydrophobic extraction, ion exchange, size exclusion (up to 100 g)	45
Separation	ST/Lab 300 peptides and proteins	45
Technology	ST/Process 200 (kg quantities)	125
Millipore	Kiloprep 250 (from 25—750 g/run)	38—55
YMC, Inc.	PLC-200 XP (500—1000 g/run)	75
Varex	PSGC-10/40	40

Costs are approximations, since a good deal of custom design is involved. A "standard" unit would probably be a rarity. Manufacturers should be contacted for pilot equipment or process evaluation. A far better cost estimate would result after some operating data were obtained.

Electrodialysis is another method for separation and purification of biological mixtures.[85] The process uses highly conductive, ion-selective membranes and low voltage current to move low mol wt organic and inorganic ions from one solution to another. While large-scale plants exist for water desalination and desalting of whey, some newer operations involve more complex separations with rather costly constituents. Proteins, peptides, enzymes, and other biological materials have been separated with a high degree of specificity. Advantages of electrodialysis are

1. No dilution of product
2. System is modular; electrical control of production rate
3. Yield is high and both high and low MW species can be desalted; very little product lost via transport or adsorption
4. Ease of scale-up
5. Low pressure operation
6. Precipitates have little effect on the overall process
7. Cleaning and sanitation are relatively simple; no other reagents or chemicals are required

Applications in bioprocessing include:

1. Removal of 99% of 2 N salt from an α-interferon eluate
2. Deionization (95%) of 1% bovine serum albumin
3. Deionization (95%) of 1-leu-try-meth-arg-phe acetate
4. 15.7 μM Converting enzyme inhibitor was deionized
5. Vasopressin was deionized (99% deionized in 1 to 5 hr)
6. Labeled angiotensin 99% deionized

Equipment has been built which can handle a batch size of 150 ℓ. With automated controls and clean-in-place capability thousands of liters per day can be processed.

A complete review of electrodialysis (ED) is available in a recent publication.[86] History of its development is given. Ion exchange or ion selective membranes are the heart of the ED system. The membranes must have qualities of mechanical strength, chemical inertness,

FIGURE 11.   Commercial electrodialysis system. (Courtesy of Ionics, Inc.)

dimensional stability, and high ion selectivity. The article details theory of operation and describes typical equipment. The basic building blocks are a membrane stack, a hydraulic system, and a rectifier as a source of direct current. Ionics, Inc. has a 30-m² ED unit (called the Bio Prep® system) which is in use for interferon purification. Another system for whey desalination processes 125,000 to 440,000 kg of raw whey per day; throughput is dependent upon deashing level (50 to 90%). Effective cell pair area is 180 m². Most food and pharmaceutical applications involve operation in a batch mode. Product composition is maintained by varying batch cycle time. With a CIP system connected to the ED unit, automated, round-the-clock operation is possible. Economics are discussed, but only for large-scale commercial applications involving brackish water or whey. For the former, operating costs are $0.26 to 0.52/m³. These are far below costs incurred with biological streams, since desalting levels would be higher and yield is far more important. For whey ash removal, production capacity is doubled when demineralization to 75% rather than 90% ash removal is sufficient. Capacity doubles again if demineralizing to only 50% is satisfactory. Operating and maintenance costs (not including depreciation) for the various cases are 11.6¢/kg product solids for 90% ash removal, 6.4¢/kg for 75%, and 3¢/kg for 50%.

ED systems are said to be especially useful for separation and purification of novel biosynthetic proteins, since apparatus is self-contained, fully automated, of sanitary construction, and can be cleaned in-place and sterilized between batches. Scale-up and determination of cost of operation are readily possible from laboratory evaluations. The Ionics Bio Prep® Mark III is shown in Figure 11. This system is a commercial unit for desalting of protein solutions. Nonionic species can be separated from ionic species at high yields and with very high purity. Operation is manual or automatic. Automatic shutdown and alarms are included and CIP systems, while optional, can be provided. Since each system is essentially a custom-designed unit and feedstream characteristics as well as desired purification determine initial and operating cost, it is difficult to generalize. However, a basic unit would start at about $400,000. A major factor in operating cost is depreciation and another critical item is energy; both are strongly dependent upon input and effluent stream characteristics.

A recently published book covers commercial-scale developments in purification of products found in fermentation broths.[87] Ultrafiltration is discussed as is HPLC, process scale chromatography, immunosorbent chromatography, and scale-up of affinity chromatography. In one effective chapter in this book, Wildfeuer[88] reviews planning and evaluation of alternatives in purification of an antibiotic (cephalosporin C). Aternatives are given with pros and cons and potential process development alternatives are reviewed. The factors that go into listing of alternatives, planning of work, and selection of unit operations are well known, but they bear repeating:

- Capital costs
- Processing costs
- Throughput requirement
- Yield potential
- Technical expertise available
- Conformity to regulatory requirements
- Waste treatment needs
- Continuous or batch processing
- Degree of automation desired or required
- Personnel safety and health

While this listing refers to purification of a specific antibiotic, it is worthy of repetition and reinforcement at every step of the project cycle.

J. Drying

Although there are cases where final product form is liquid, most often a dried product is desired for use in formulation, whether as a powder, a slurry, or in a standardized solution. The major categories which impact drying — removal of a liquid from a purified product stream by evaporation — are properties of the solid-liquid solution or mixture, properties of the resultant solid, properties of the environment for drying, and heat transfer characteristics. Since there are many methods of gas-fluid-solid contacting, with and without application of heat, there are many physical arrangements to effect drying.

In biotechnology applications, many products are heat labile. Therefore, mild operating conditions or short residence times (or a combination of both) are employed. Flexibility is once again desirable, as is ease of cleaning, so that different products may be dried in a campaigned manner. At this stage of processing, any loss, however slight, will be costly, since purity is highest and product has accumulated almost all absorbed costs of manufacture. Contamination is never a permissable option, but at earlier stages of processing there remain purification steps that would remove foreign materials. The only steps following drying are normally standardizaiton and sterilization (which may involve separation).

Dryers can be batch or continous and fall into two main categories: direct drying (direct contact between wet stream and gas, also called convective drying) and indirect drying (vaporized liquid removed independently of heating medium, also called conductive or contact drying). Drying can also be accomplished by means of radiation or via high frequency radiation; industrial applications are more limited. There are many characteristics which must be reviewed in dryer selection:

General	Biological/pharmaceutical
Physical properties (wet/dry)	Corrosiveness
Explosive nature	Toxicity/immunology

Abrasiveness	Flammability
Shrinkage	Particle size
Solvent/dust recovery	Heat stability
Dryer residence time	Open/enclosed (sensitivity)
Predryer treatment	Contamination
Product adhesion	Bulk density
Product removal	Inert atmosphere
Available space	Turnaround (alternate products)
Installation/operating costs	GMP requirements
Feed rate	Solubility/compressibility
Angle of repose (certain designs)	
Viscosity	
Hygroscopicity of product	

One general-use drying technique found in many pharmaceutical operations is tray drying, usually under vacuum. There is normally provision for heating each shelf or region of the dryer (with a reasonable temperature range). Programmed heating is possible so that less stringent conditions can be applied over the drying cycle. The chamber itself (cylindrical or rectangular) can be evacuated so that drying is affected at reduced heat input. The positives for such equipment are useful for heat-sensitive materials, small volumes can be dried, and general applicability. Negatives are high labor cost, high unit cost of drying, and relatively large space requirement per unit of product.

A specialized form of vacuum drying is lyophilization or freeze drying. In large-scale lyophilization, materials to be dried are again placed into trays with a relatively low depth of liquid. Trays are placed on shelves which are refrigerated rather than heated (although both functions may be incorporated). The liquid is frozen and then water (if aqueous) is removed by sublimation under vacuum. The drying technique is costly (noted above), but also very useful for sensitive materials. The aim of the operation is to maintain ice temperature at a value just below melting, while the product surface is heated to the highest temperature it can withstand safely. Vapor removal is achieved by means of vacuum pumps and condensers. Location is important to prevent undesirable pressure drops and reduction in sublimation rate. Freeze drying is an expensive operation and is usually employed when other means prove unsuitable. The process is suitable for many pharmaceuticals and fine chemicals.

Various types of dryers utilize a means of suspending product in hot air (a spray dryer is one such type). A pneumatic dryer is basically a long duct into which liquid mixture is introduced along with high velocity air. In a ring dryer, controlled recirculation is used to dry product; however, the principle of suspension in conveying-drying air is retained. Product may be supported in the form of a bed of varying density by drying air. In such a design (the fluidized bed dryer) conveying of product by air is undesirable. In appearance, the supported bed looks like a boiling liquid. Escaping gas gives a "bubble burst" surface appearance. The intimate gas-solid contact results in efficient drying. Moist material is fed into one end of the machine (length is a variable as is air temperature and air velocity) and dry product is discharged at the opposite end of the bed.

In many cases, pneumatic dryers are used for heat-sensitive, explosive, flammable, or easily oxidized products, since residence time is short. Furthermore, there is little or no product holdup which allows for ease of cleaning and multiproduct use. A recent report describes a system designed specifically for fine chemical and biotechnology applications.[89] A slurry is introduced into the system, and solids are separated in a perforated basket centrifuge and dried in a pneumatic conveying dryer. The system is totally enclosed, which means that contamination (both inward to product and outward to cause potential personnel exposure) is eliminated or minimized. Drying times vary from 1 to 60 sec. The advantages claimed for pneumatic conveying drying are

1. Fast cycles; minimum residence time
2. High thermal efficiency
3. Automatic solids transport
4. Small size relative to other dryers
5. Ease of control
6. High capacity

Pilot equipment is skid mounted and the small unit, if dedicated to a single product, could product 20 tp 200,000 kg/year. It is a relatively simple matter to inert the entire apparatus. Solvent recovery is possible. A wide range of pharmaceuticals and fine chemicals has been processed (examples are penicillin, streptomycin, vitamins, cephalosporins, calcium acetate, various herbicides). Sterile and particle-free processing is achievable. Typical cycle time is 5 to 29 min, with recoveries of 2 to 16 kg/cycle claimed. The pilot unit cost is approximately $350,000.

The author (W. K. Pearson) has kindly supplied additional information on the system. All materials in contact with process streams are constructed of 316 stainless steel with a 4-μm finish.

The smallest commercial-sized unit has about eight times the capacity of the pilot plant. An additional centrifuge doubles system capacity. More centrifuges can be added to increase output; therefore, investment does not increase according to the well-known exponential formulas. Budget pricing for these units (mid-1986) are

Cost ($000)

Pilot unit	350
8× Increase	650
16×	800
24×	950

Operating labor is estimated at $^1/_2$ man per shift regardless of capacity. For a unit designed to produce 20 million lb/year of the analgesic, acetaminophen (APAP), utility requirements are

Air	20 ACFM at 80 psig
Steam	435 lb/hr at 150 psig
Electricity	105 kW hr peak

Spray drying is a convenient method for many materials. A feed solution or mixture is atomized into a specially designed chamber where air (usually, but not always, heated) is introduced co- or countercurrently. Atomization is achieved by means of a high pressure pump and nozzle system or by feeding onto a high speed rotating disk. As droplets are dried they move in a swirling motion to the bottom of the dryer or into separating cyclones. Dry product settles and is separated from conveying air by means of special discharge valves. Resulting product usually takes the form of very small spheres. Thermal efficiency is poor, a complex filtration system is usually used for exit air (baghouse), and large volumes of exhaust air may present an environmental problem. However, contact time is very short; it may be a matter of seconds. Further, the design of the dryer allows for flexibility. Heated gas may be introduced at many locations, thus, changing the internal flow pattern and

Table 21
TOWER DRYER CAPACITY VS. COST

Water removal rate (lb/hr)	Power requirement (kW)	Heating requirement (million BTU/hr)	Price ($000)
1000	55	2.2	350
1000ᵃ	50	2.2	400
2000	90	4.2	570
2000ᵃ	80	4.2	650
4000	170	8.0	730
4000ᵃ	160	8.0	800
6000	270	11.0	790
6000ᵃ	250	11.0	850

Note: Add 15% to prices for installation. Heat recovery is not included (savings of up to 25% in heating requirement are possible).

ᵃ With baghouse.

temperature profile. Cool gas can be introduced at separatory cyclones (dehumidification is essential). The major design input is to achieve particles of a specified size with a relatively narrow size distribution. This not only insures proper drying of the "largest" particle, but prevents overheating of solid particles from the more rapid drying of "small" liquid droplets.

It should be recognized that spray drying is rarely, if ever, cost competitive with the more "common" dewatering-drying options. Removal of particles from effluent gas (stripper, baghouse) is costly and heat recovery is complex and costly, even if feasible. Spray drying offers certain advantages and the additional costs must be justified on the basis of some improvements in product characteristics. Examples of such improvements are particle size, stability through drying step, color, flavor, retention of activity, or capability of encapsulation. Added auxiliary equipment includes air filters, insulation, dehumidification systems, cooling conveyors, and special alarms (or shutoffs) for high-temperature or feed interruption. At minimum, inlet and outlet temperatures should be known, pressure and pressure drops should be monitored, feed pressure and rate should be measured, and drive motors' power draw (high pressure pump or disk drive) should be indicated.

One supplier of tower dryers (Damrow Company) has presented budget estimates for various capacities.[90] These data are approximate and were given in early 1986; the supplier should be contacted for up-to-date estimates. Table 21 shows water removal rate, heat and power requirements, and approximate equipment price.

Other types of dryers include

- Conveyor: perforated band conveys product
- Roller: liquid fed to one or two heated metal rolls, knife edge removed dried film
- Rotary: rotating hollow shell with heated gas fed co- or countercurrently
- Agitated: solids movement within stationary housing is provided by an internal agitator

A rotary dryer consists of a metal cylinder with or without flights. Heating may be direct (direct exchange between flowing gas and solids to be dried) or indirect (annular space in cylinder walls or one or more rows of metal tubes within cylinder to convey heating medium). Obviously, product should be fairly free flowing and granular. Provision for partial product recycle can improve product characteristics. Separation and/or recovery is essential when a solvent other than water is involved or when solid particle entrainment occurs. Pricing for warm air direct-heated cocurrent rotary dryers is given in Perry as of early 1982.[91] System includes finned air heaters, transition piece, drive, dryer, product collector, duct, and fan.

The smallest unit (1.22 × 7.62 m) will evaporate 136.1 kg/hr and would cost $150,000 (304 stainless, FOB Chicago). The largest unit detailed (3.05 × 16.77 m) will evaporate 861.8 kg/hr and would cost $480,000. Process assumes a heat-sensitive solid with maximum temperature of 65°C.

Agitated dryers have been used in pharmaceutical applications; vacuum can be readily applied to conventionally designed systems. A turbo-tray dryer is a special design that permits variation in temperature and residence time while occupying limited floor space. Further, many different materials, from fine powders to thick slurries, can be handled. One manufacturer (Wyssmont) has presented data on drying of an antibiotic filter cake.[92] The interior is stainless steel, height is 7.6 m, diameter is 6 m, and it is steam heated with an evaporation rate of 910 kg/hr and a dried product rate of 2400 kg/hr. Unit was field erected and cost (1980) was $390,000.

Reviews of various drying operations are available. Forrest reviews dryers that are applicable to biological materials, including foods.[93] Theory and fundamentals of drying are also discussed. One point is made that is often overlooked. The most economic moisture removal method is by mechanical expression and this should be the first step considered in any drying operation. This step can be broadened to settling and centrifugation (where a separate phase exists); recycle of mother liquor or supernatant will retain product for subsequent recovery.

V. PROCESS EVALUATION, PFD, PID

The major steps in fermentation are medium preparation and sterilization, inoculum handling and development, air input, feed preparation and monitoring, process monitoring and control, sterility maintenance, and harvest. The extraction train will be product specific but involves impurity and water removal in different unit operations and/or processes. Detailed specifications which are formulated in engineering drawings and diagrams will insure correct sequencing and capability to achieve planned output.

Material balances become the basis for equipment selection, valve, piping, instrumentation, and storage design; equally important are expected rate and yield which determine true product cost. Mass and energy balances determine sizing of heat exchangers, utilities, cooling media and transport, evaporators, drying equipment, as well as sterilization needs. There is a more prosaic reason for developing both mass and component balances. Any effluent stream, whether gaseous, liquid, or solid, must be disposed of in some manner. Atmospheric discharge of spent air by itself presents no problem, but spent air containing a fractional percent of H_2S or NO_x cannot be discharged without treatment. There are limits being placed on water discharges where no limits existed before. Water containing any measurable solids, BOD, or ammonia must be carefully monitored and must be treated; this means money and may involve major investment. Furthermore, some unexpected disappearance (or appearance) of a stream or material may indicate some break in a tank or line; results can be not only costly, but catastrophic, depending upon composition.

Substrate is utilized for cell growth, cell maintenance, and product formation. References are available which discuss cell development and biosynthetic product formation.[94,95] A carbon balance is useful not only for the fermentation step but also in subsequent purification. The fate of specific carbon atoms elucidates mechanism as well as aids in determining efficiency. A carbon balance developed during lab and pilot stages is a useful scale-up tool and can direct process development. Differences noted at production scale may point to direction for improvement procedures. Similarly, a nitrogen balance will help in determining efficiency and may be modified with changes in the nitrogen source. The nature of the desired product, as well as the nature of impurities, determines potential treatment steps. Stability of the many bioactive compounds is also a determinant. For example, sale of dried cells or a liquid side stream that contains traces of an antibiotic or of a growth hormone would not be permitted. Destruction by chemical, physical, or biological means would be required. Flow diagrams will also show recycle streams; cost reduction (as with condensate recycle) is possible with appropriate stream selection with or without partial purification.

A process flow sheet (or its expansions) includes important design information. This sheet leads to the overall material balance for the process. Chemical and physical data (including composition) for each stream in the process should be tabulated; special importance should be given to viscosity, corrosivity, temperature, pressure, pH, and other variables related to design. Along with these data, energy inputs and energy removals should be shown; energy balances will lead to design of appropriate heat exchangers, evaporation or distillation components, and drying equipment.

One common piece of equipment is a heat exchanger. A single heat exchanger can have approximately 100 items of information specified prior to reaching a design and quotation; both process and heating medium side must be included. Certain variables include:

Material balance	Percent solids
Flow rate	Corrosion allowances
Pressure/pressure drop	Material of construction
Temperature (in/out)	Regulatory (code) requirements
Heating or cooling medium	Noncondensables
Specific heat	Thermal conductivity
Specific gravity	Latent heat
Viscosity	Cleaning methods/frequency
Molecular weights	Design temperature/pressure
Contact surfaces (gaskets)	Fouling factors

While all data are not shown on a process flow diagram, pertinent data must be included in the design package as an accompanying bound volume. There may be one design file or book solely concerned with heat exchanger design, while another might cover pump design. Where generalized information and specifications can be presented in a package, such collation should be made.

Engineering flow sheets can be considered expansions of the process flow sheet. These are grouped and numbered to give the complete design package. Equipment is shown in association with all related piping, valves, and instruments. Main process flows are shown, relief devices are detailed, and sizes are specified. Flowsheets may be further subdivided, as:

Process flow	Steam piping
Electrical system	Lighting
Instrumentation	Starters
Heating/air conditioning	Air piping/vacuum system
Water/cooling water	Sewer streams and connections

Piping and instrumentation diagrams should show issue (sequence number or letter), date of issue, and required approvals. Revisions can become complex and the sequence that led to the ultimate design is important. The need for approvals is obvious; nonetheless, it is sometimes overlooked. Not surprisingly, ultimate construction is not exactly what everyone agreed upon and costly revision is the usual result. A PID for a fermentor will show pipe sizing, valve type and sizing, relief devices, instrumentation including indicators, final control elements, service for each line, leads into and out of drawing (to associated equipment drawings), special notes for construction details and materials of construction, special mounting methods, interconnections, pressure/vacuum ratings, agitator and filter ratings, alarm points (preferably with alarm limits), and general orientation relative to the fermentor. It is a good idea to follow all process steps in the vessel (wash, sterilize, charge, inoculate, feed, harvest) while following flows and process control loops shown on the PID. Required readings, control inputs and outputs, and system response (theoretical) can be followed. Any line that is extraneous, or any line that is missing, can be identified by a forceful correlation of operating instructions with the PID. A typical fermentor PID is given in Appendix B.

It should be remembered that the engineering drawings of all types will remain at the plant site. These drawings will serve not only as an historical record (modification, repair, replacement, upgrading), but will serve as a training vehicle. The care and attention they receive in planning and execution will pay dividends on start-up and in the course of routine operations. Should some nonroutine or unexpected event occur, the drawings will prove invaluable. Since design attention and design presentation are important, updating of drawings should be stressed. Numerous changes will be made during start-up and other changes will be made, not only in the course of running the plant, but during scheduled shutdown periods (as during annual shutdown). Very often these changes are not incorporated into the engineering drawing file. At best, such neglect will mean inconvenience during some emergency; at worst, a safety problem can result. One example will suffice. Changes to high voltage circuitry or changes in load to a major motor control center that are not clearly and carefully documented can put personnel at risk during unscheduled maintenance and repair.

Utilities should be reviewed as carefully as process equipment. There may be a tendency to leave utilities planning to experts in their design and presume that appropriate drawings have been rigorously reviewed. This tendency should be fought and overcome. An in-depth knowledge of the process is needed to appreciate impact of a failure or breakdown in a service or the need for redundancy in supply. Utilities in a fermentation plant may include:

Cooling water and return	Plant air (sterilization)
Process water (treatment, if any)	Instrument air
Steam distribution	Refrigeration
Condensate distribution	Cooling tower system
Well water flow	Brine system
Chemical treatments	Electrical system
Sanitary piping	Waste flows (treatment)
Vacuum system	Inert gas system

If there are in-place cleaning systems (especially recirculating), these units should be included in separate drawings. Plumbing and HVAC systems should be detailed separately. Air movement is important not only in office and laboratory areas, but also in large plant areas. Stagnation or condensation problems can be solved after the fact, but cost is far higher.

Taking just one utility service for some detailing, we can see that a rather large planning effort is needed for even a small plant. While other utilities must be planned with other functions and services in mind, a generalized requirement for each utility can be deduced from specific drawings for design of a typical electrical distribution system:

1. Site plan with major details needed so that distribution achieved at minimum cost; interferences should be considered and appropriate corrections made
2. Single line diagram

3. Lighting, including distribution, types of lights, indoor and outdoor requirements
4. Individual office and plant areas
5. Laboratory power
6. Motor control center (MCC) schedule
7. Conduit and panel details
8. Feeder and busway details
9. Total and individual metering (if desired)

Tie-ins of all services to all process equipment must be shown on engineering drawings. Branching and tie-ins should be planned to minimize disruption, should some major or even minor failure in the utilities area occur. Alternate supply of essential services should be planned. The need for stand-by services should be determined in the design phase and provision made for such installation. This is especially true for needs during a catastrophic failure or loss (fire suppression, electrical services, communication). A crystallizer might have stand-by provision for a fuel-powered drive or for air suspension (fuel-powered compressor) in case of a major power outage.

As can be imagined, computer-aided design has had a marked impact on plant design. Process simulation or flowsheeting systems are being improved on almost a daily basis. Most large computer manufacturers offer their own systems; however, specialized companies and centers have grown to supply computer aids in the CAD field. The programs can provide heat, material, and component balancing as well as sizing and costing calculations. If a selection of a process is made, a list of individual components (tanks, agitators, pumps, etc.) can be printed. With appropriate data files, battery limits investment can be presented for any selected alternative. A recent review article discusses a number of specialized companies or projects concerning process flowsheeting.[96] Some of the systems and companies are DESIGN 2000 (Chem Share Corp.), PROCESS (Simulation Sciences, Inc.), CONCEPT (CAD Center, England), ASPEN PLUS (Aspen Technology, Inc.). While earlier designs (meaning those of two or three decades ago) stressed oil, gas, and petrochemical applications, the advanced systems have been applied to biochemical processes, inorganic chemicals, pulp and paper industries, and mineral processing. Optimization capability is becoming a more common accompaniment. Process flowsheeting has become an essential part of an integrated approach to computer-aided process plant design.

A very interesting article is worth consulting for modern computer-aided design as applied to food processing. While the field is a dedicated one, there are many overlaps and applications to design and operation of a biotechnology plant.[97] The analysis is applied to steady-state processes (as corn wet milling or whey concentration and drying) or single-product batch/semicontinuous operations (as cheese production) or multiproduct batch/semicontinuous operations (as soups or confectionary products). It is obvious that sterile fermentation processing falls into either the single or multiproduct batch/semicontinuous mode. Design and equipment selection must be appropriate for dynamic, unsteady-state operations where multiple-use process lines and equipment would be most economic. Problems of integrated design and operation become larger in scope and computational complexity. The reasons for this degree of complexity have been noted earlier: need for automated media preparation and batch sequencing (including sterilization), in-process inventory management (with appropriate conditions of storage), complex materials requirement planning and scheduling, and associated allocation of labor and energy. An added burden in pharmaceutical processing is the need for conformance to GMP and regulatory requirements. Records (whether paper or electronic) must be suitable for the intended product distribution and use.

Okos and Reklaitis[97] describe and list 12 references for computer-aided methodology for design and operation of noncontinuous food processes. Simulation packages that are oriented toward noncontinuous operations are discussed. Examples are given in process synthesis (review of various processing routes with estimation of capital and operating costs), material and energy balancing, preliminary design (including modular capital cost), and process simulation (emphasizing the BOSS package which simulates a multiproduct batch/semicontinuous process with intermediate storage). A prototype integrated work station is described which draws upon developments in data base management, interactive computing, and computer graphics. Such a work station allows interactive flowsheet drawing, material and energy balancing, equipment sizing, and equipment costing. A progressive flowsheet is very useful, because selected pieces of equipment can be removed and others inserted without starting from scratch. Several alternatives can be tested to find an optimal solution. It would be a relatively simple matter to move from process design to manufacturing cost estimation and from there to various business routines to determine cash flow and return on investment.

Speciality chemical production involves a plant divided into numerous processing areas where a number of products can be made. Batch operations are common. Complex reaction paths with long reaction times may be the norm. Solids handling is often involved. There are control difficulties and uncertainties along with feedstock variations. Delays due to operator unavailability occur with some frequency. Other unpredictable variations occur and system bottlenecks are not always apparent. Those familiar with fermentation operations will recognize all these symptoms and will probably feel that any analysis of systems with those characteristics would be applicable to their situation. A simulation method has been developed that involves capacity planning and determines impact of selected process modifications on overall plant productivity.[98] Some model processes are described involving many different steps, with each production stage having certain items of equipment. For each simulation, the user must specify: number of reactors, crystallizers, stills, centrifuges, and operators as well as capacities of each unit and all intermediate storage tanks. Also to be specified are processing time distributions (including cleaning and maintenance), product specifications, process conditions, waste disposal rates, labor for each step, alarm conditions, and initial status. Methodology was developed after plant operating data were collected and organized. Productivity changes and estimates were developed which can be used to determine cost-to-benefit ratios for a selected modification. Such a program would be useful in design phases as well as in upgrading projects in an existing plant.

One commercial system for process evaluation is supplied by Enyedy.[99] The concept of process evaluation has as its objectives:

- To determine venture profitability
- To justify funding for a project
- To compare process alternatives
- To select among processes
- To investigate competitive processes
- To prepare preliminary estimates for new or modified plants
- To select the best among alternate investments

Process evaluation should include economics as early as possible in the design sequence. The steps for analysis of any process should include:

- Identification of key process steps
- Preliminary flowsheet with identification of major unit operations and associated equipment
- Material and energy balances
- Sizing (selection) of equipment
- Costing of equipment
- Estimate of total plant estimate
- Estimate of manufacturing cost
- Calculation of selected profitability indicators

Enyedy (company is PDQS, Inc.) has devised a combined program that incorporates chemical engineering practice as well as cost engineering methods. This process evaluation tool is called PROVES. The program is an integrated package with four parts that is especially applicable to conceptual design. Starting with a prospective flowsheet, a process simulator program calculates the mass balance, an equipment sizing and costing section selects captial items and costs, an investment estimator results in preliminary ($+30$, -15%) estimates, and detailed operating and profitability determinations are made.

The process flowsheet for mathematical simulation can be prepared by the user or sophisticated data bases (described above) can be used. In many instances energy balances can be projected. In the PROVES subsystem (called MODEL), nine basic operations are available: mixing, splitting, subtraction (two types), separation (two types), reaction, addition, and equilibrium. MODEL can simulate batch and cyclic processes (can handle 99 streams, 30 components, 80 operations). The next subroutine (SCOPE) sizes and costs major equipment items from the flowsheet. Major equipment classes are

Absorbers	Air exchangers
Bins, silos	Compressors
Crushers	Distillation columns
Dryers	Extraction columns
Filters	Fired heaters
Heat exchangers	Mixers
Pressure vessels	Pumps
Reactors	Batch reactors
Gas scrubbers	Tanks

Also estimated are utilities requirements for each major piece of equipment (electricity, water, steam, heat transfer fluid, refrigerant, and fuel).

The total plant investment or fixed capital cost is derived from cost of major equipment using published and modified factors and ratios. Plant investment includes:

> Direct items
>> Process equipment
>> Process utilities
>> Nonprocess buildings
> Allocated facilities
>> Supporting utilities
>> General facilities

This portion of the program is called INVEST. The final element of the overall program is EFFECT which incorporates utilities costs, investment cost and other economic data to estimate operating cost, working capital, sales, profitability, and sensitivity analyses. The PROVES program was loaded onto a UNIVAC 1108 in 1973 and was made available to users as remote batch. In 1979, PROVES was rewritten and enhanced for a time-sharing system. In 1984, it was again rewritten for micros and is called PROVES IV. It is operable on an IBM PC™ DOS and the minimum hardware requirement is a 512 K memory IBM-XT™. The program is available through a consulting license arrangement. Cost is $15,000 over a 3-year payment schedule. Continuation can be provided for two more years for a total of $10,000 additional over that time.[100]

Table 22
SELECTED COST INDICES (CONSTRUCTION COST)

Year	Chemical equipment	Process plants	Petroleum refinery	General construction
1975	444	182	200	207
1976	472	192	215	224
1977	505	204	223	241
1978	545	219	244	259
1979	599	239	260	281
1980	660	261	286	301
1981	721	297	315	329
1982	746	314	340	356
1983	761	317	357	377
1984	780	323	369	386
1985	790	325	374	389
Base	1926 = 100 Marshall and Swift	1957—59 = 100 *Chem. Eng.* plant index	1967 = 100 Nelson refinery index	1967 = 100 *Eng. News Rec.*

VI. CAPITAL COST

The total cost of capital for a project consists of these elements:

1. Fixed capital costs
2. Working capital costs
3. Cost of land and other nondepreciable costs

The fixed capital investment is regarded as the capital required to provide (purchase and install) all the depreciable facilities for required production.

There are useful references for both an overview and detail in preparing capital cost estimates. There is no need to summarize that information, but certain selected areas of interest will be covered. The complexity of the task can be gauged by the number of items covered in a checklist (Reference 101, Table 25-43).

Simple cost indices can be used for rough cost estimates if costs are known for some base period. It must be recognized that commodity prices change due to inflationary and other factors, equipment costs may increase at a different rate, and construction cost may increase at yet a third rate. Composite indices may be derived, but these only simplify calculations without introducing any greater accuracy. Examples of selected construction cost indices for the U.S. are given in Table 22. The earlier data are taken from Perry (Table 25-41) and updated from the respective sources.

Estimates based on ratios can be prepared without design work, a flow diagram, or process flow information. They are based on historical data and are corrected for time and plant size. Annual sales may be included, but this would have little validity for many biotechnology products. Ratio methods are useful for relatively large volume commodities, but even here potential error is large.

Exponential methods use capacity-ratio exponents, again based on published information. There is little published information on fermentation plants; what is published is very product specific and may or may not include exaggeration related to expected process development. In other words, exponential methods will have greater validity for products such as styrene, chlorine and caustic, acetaldehyde, or ammonia as compared to penicillin or vitamin B_{12}. What may be of greater use are exponential relationships for specific pieces of equipment. It is possible to estimate single equipment cost over a wide range of capacity or throughput, since a reasonable amount of published information is available. Some suppliers can provide order-of-magnitude estimates for differing sizes. Within limits, interpolation or extrapolation is a reasonable procedure.

Equipment cost functions over a range of sizes are presented in a number of publications. One such table lists 13 classes of equipment and uses the following formula[102]:

$$\text{Capital cost} = \left(\frac{\text{MSI}_{now}}{\text{MSI}_{base}}\right)(C_{base})\left(\frac{A_{now}}{A_{base}}\right)^{f}(f_{bm})$$

where C_{base} = base cost in dollars: MSI_{base} = Marshall-Swift index, base; MSI_{now} = Marshall-Swift index, current; A_{base} = base equipment size, convenient units; A_{now} = current equipment size, same convenient units; f = an exponential cost factor; and f_{bm} = conversion factor for base cost to module cost.

Let us take the example of a cation exchange column:

$$\text{MSI}_{base} = 190 \quad C_{base} = \$25,800$$

$$A_{base} = 100 \quad f = 0.7253$$

$$f_{bm} = 1.28$$

and the size range given for "safe" approximation is 5 to 20 ft³. Similarly, for a spray dryer:

$$\text{MSI}_{base} = 185 \quad C_{base} = \$51,922$$

$$A_{base} = 1000 \quad f = 0.705$$

$$f_{bm} = 1.1$$

and the size range for use of the correlation is 100 to 10,000 lb/hr. If the correlation were used to cost a dryer having a water removal rate of 4000 lb/hr, 1985 cost would be $648,000. Elsewhere in this book, a tower dryer builder estimated cost at $730,000. A difference of about 11% is within the expected error of such a rapid estimator. Cost correlations for equipment are given by Woods.[103] Twelve pages give a rather complete listing of process equipment. All prices refer to North America for mid-1970 corresponding to a Marshall-Stevens index of 300. Most prices are for carbon steel with some quotations noted for stainless construction. The table lists size, unit of measure, cost (U.S. dollars), range of size, and the exponent for scale-up or scale-down. In many cases, percentage error is shown. Taking the same spray dryer as discussed above (and a stainless to carbon steel cost increment of 2), one calculates a FOB cost of $679,000. Once again, agreement to a published quote is not too bad. The listing covers equipment that might not be found in the more conventional compilations. Of interest to a designer of bioprocess plants are data on:

Activated sludge units	Electrodialysis
Various crystallizers	Podbielniak extractors
Dialysis	Ion exchange systems

A more limited list of process equipment with exponents for capacity changes is given by Perry et al.[101] (Table 25-49).

Capital cost estimation (as part of the financial decision-making process) is also covered by Holland et al.[104] This review discusses the various types of estimates with associated probable accuracy. An order of magnitude or ratio estimate has a probable accuracy of $\pm 30\%$ while a firm or detailed estimate has a probable accuracy of $\pm 5\%$. There is a short list of factors which tend to increase capital cost. These are

1. Overprovision for safety
2. Overprovision for stand-by equipment
3. Unnecessarily robust supporting structures
4. Enhanced enclosures or unneeded enclosures
5. Overdesign in materials of construction
6. Inclusion of nonstandard size or type of equipment

One could add overdesign for environmental control and overinstrumentation. However, there is a rather thin line one must follow in design of a fermentation facility, especially one where containment is an issue. A certain amount of overdesign might be construed as prudent. Ratio methods are discussed and capital cost determination based on design is reviewed. Examples are given.

One rapid estimation method involves use of factors. Factor methods include materials, labor, and various overhead components. The factor method itself has many degrees of complexity. Normally, the base cost of equipment is needed so that a flow diagram and mass and energy balances are required. Other base costs, or actual costs, can be incorporated so that more accurate weighing of calculated values is possible. Correction for geographic location can be included. In costing of equipment, it is simple (but erroneous) to neglect standardization of the base estimate. Equipment cost can be FOB (on delivery vehicle from point of manufacture, port of entry, or other defined location), DEL (delivered but not unloaded at site of location of use), or installed. Factoring methods can begin with FOB costs and incorporate the others by estimation.

A series of three articles was published in 1985 which updates cost factors for chemical plant capital investment.[105] The first paper discusses process equipment and an eight-step procedure which includes (in a sequential fashion):

1. Battery-limits installed equipment
2. Plant direct cost
3. Total direct cost
4. Construction cost
5. Depreciable capital, excluding contingency
6. Depreciable capital
7. Fixed capital
8. Total capital (includes working capital and start-up)

Selected equipment costs (plots of cost vs. area, power, capacity, or size) are given. The second paper gives commodity (bulk) material cost factors for 29 different process equipment items. The material costs include: concrete foundations, piping, steel supports, instrumentation, insulation, electrical, and painting and costs are given as dollars for material per $100 of process equipment. For a vertical pressure vessel, total "material cost" would be $108/$100 of equipment cost; an agitator would add $29/100. A long-tube or forced circulation evaporator has a cost factor of $141/100; it is the highest such total for all the items shown. Other factors are described and direct and indirect cost factors (cost base and range of factor as percent of cost base) are listed. An example is given. The third and last article discusses installation-labor cost factors for 25 categories of equipment with many subgroups. Various alternatives are shown and variation with escalation and labor productivity is also shown.

There are also more concise (which also equates to more assumptions) compilations which can be used to convert delivered equipment costs into fixed capital investment. Each item of the construction program (including field expense, engineering, contractor's fees, contingency) is detailed and the total fixed capital investment estimated. Table 25-50 in Perry gives such a table.[101] For delivered equipment cost of 1.00, a grass roots solids-fluid processing plant would have a fixed capital investment of 4.27. For a battery limit installation (same base), the total fixed capital investment would be 4.17. Similarly, factors or percentages of various base costs which can be used to estimate cost for auxiliary facilities as a percentage of total installed plant cost are given by Perry (Table 25-60). The table refers to grass roots plants and ''large'' additions. The lowest median percentage is 0.2 for yard and fence lighting and the highest median percentage is 5.0 for auxiliary buildings.

Woods[106] presents cost correlations for auxiliaries. Frame of reference is mid-1970; a base size and cost are given as well as a range of capacities, exponent for scale-up or -down, and percentage error. Various packages for steam generation are shown as are alternatives given for refrigeration and waste disposal.

There are also procedures for adjusting for cost differentials within the U.S. and for construction in overseas locations. The factors to be reviewed, determined, or estimated prior to selection of a foreign manufacturing site are listed in Perry (Table 25-64).

Continuing with the example of citric acid, it is possible to combine cost indeces, equipment selection, and an older estimate[107] to make some reasonable estimates of capital investment for a plant designed to produce 40 million lb/year (18.18 × 10⁶ kg/year) of citric acid. Process flow has been given earlier. The first step in capital estimation is preparation of a list of process equipment. An example is given in Table 23. Utility requirements are also shown. The complexity of the equipment list should correlate with the degree of accuracy required. When a definitive estimate is reached, the equipment summary should be equally ''definitive''. The fermentation section has a continuous sterilizer, four 15,000-gal seed fermentors, and eight 100,000-gal production fermentors.

Some key assumptions made prior to equipment selection are

Harvest volume	75,000 gal, 283.9 kℓ
Citric acid content	15% w/v
Extraction efficiency	95% (42,585 kg × 0.95 = 40,456 kg/batch)
Cycle	5 days/batch + 10-hr turnaround = 130 hr
Hr/year	8160 (8160/8760 = 93.2%)
Fermentor batches/year	62.77 batches/fermentor
Total batches required	449 batches/year
Fermentation output	
(Product)	53,476 kg citric/day (× 340 = 18.18 × 10⁶ kg/year)
(Broth)	56,290 kg citric/day
Volume broth/day	469 kℓ/day
Overall yield	2.36 lb molasses (49.5% sugar)/lb citric

Using published sources as described above plus correction for time, for volume (or area), and for material of construction, one calculates a cost of $50.97 million (say 51) for a grassroots plant. This figure excludes auxiliaries, but includes listed utilities (no electrical power generation on-site), tankage, and pumps. If a figure of 20% is added for general service auxiliaries, total capital investment would be $61.2 million. If the data of Fong are

Table 23
CITRIC ACID — PROCESS EQUIPMENT SUMMARY

	No.	Area or vol (each)	
Fermentors	8	100,000 gal	Agitated, internal coils
Seed fermentors	4	15,000 gal	Agitated, internal coils
Continuous sterilizer	1	15,000 GPH	With heat recovery
Harvest tank	1	150,000 gal	Clean-in-place
Rotary vacuum filters	3	400 ft^2	One standby (swing)
Precpitation tanks	2	10,000 gal	Agitated, internal coils
Rotary vacuum filters	2	200 ft^2	
Acidulation tanks	2	6,000 gal	
Rotary vacuum filters	2	100 ft^2	CaSO$_4$ removal
Evaporator feed tanks	2	75,000 gal	Agitated
Evaporator	1	4 Effect	5,000 GPH water removal
Concentrate tanks	2	75,000 gal	Agitated
Ion exchange columns	2	200 ft^3	Carbon steel, rubber lined
Carbon decolorizaiton	1	200 ft^3	Carbon or ion exchange
Filters	2	6,000 GPH	Automatic, leaf type
Crystallizer	1		Agitated, evaporative
Centrifuge	3		Automatic discharge
Mother liquor tanks	1	20,000 gal	Agitated
Dryer	1	6 × 40 ft	Indirect heat
Storage bins	2	60,000 lb	
Classification	1		Continuous screening
Packaging line	1		150,000 lb/day rate
Molasses storage	2	1.5 × 10^4 tons	Carbon steel
Media make-up	1	80,000 gal	Concentrated media, then water
Hydrated lime feed	1	6,000 gal	Automatic load/feeder
Lime storage silo	1	50,000 lb	Live bottom bin
Sulfuric acid storage	1	50,000 gal	
Air compressors	4	6,000 CFM	Coarse filtration, in/out

		Average	Peak
Cooling water	GPM	250	350
Refrigeration	tons	150	200
Compressed air	SCFM	18,000	24,000
Process water	GPM	120	150
Electricity	kW	5,000	5,500
Steam (250 psi)	lb/hr	60,000	66,000

corrected for volume and time, the "battery limits plus utilities cost" becomes $43.85 million (say 44). Addition of general service auxiliaries at 20% (Fong uses 15%) brings the total to $52.8 million. Even if this figure were correct and the higher figure were in error, there is only a 15.9% "error" in the higher estimate. At this early stage in the cycle, it would probably be prudent to take an average figure for estimation purposes. As actual estimates or bids from suppliers come in and as the process becomes more firm, it will be possible to improve the initial value. Decisions will be made which will either raise or lower the initial estimate. Examples of such decisions are use of used equipment in selected areas, modification in materials of construction, change in percent running time or yield, and changes in utility requirements as determined on scale-up or full-scale testing.

Citric acid is a product of fermentation, but it is also a commodity chemical. As such, the major part of the investment relates to the fermentation section of the plant. This would be true for biosynthesis and purification of monosodium glutamate, lysine, many antibiotics, and vitamins and other large volume bioproducts. As has been noted, this situation will almost certainly reverse itself in the case of a new generation, high activity bioproducts. One fermentator (even one of 10 to 20 kℓ) might supply a full year's production of a novel compound; complex and costly purification equipment may be needed. Investment in extraction/purification may be considerably higher than that required for fermentation. Published information, however, tends to center upon commodity products. A judicious approach to published data, specialty purifications, and related cost data (some of which are detailed in Section IV), and supplier bids and quotes will allow the scientist to synthesize a rational cost estimate based upon the selected fermentation production process.

VII. OPERATING COST

There are many similarities in costing of a new process in a new facility or of a new process in an existing facility. It is obvious that costing in an existing plant is simplified by a history of operation and cost trending lines coupled with experience. It must be remembered that a fermentation plant is costly in real dollar terms and probably very costly when compared to volume output. Since this is the case, efforts will be made to "fit" novel processes into an existing plant. The greatest uncertainty will be present when a new bioactive product is to be made in an entirely new plant.

A number of references exist where costing is emphasized.[108-111] These should be consulted as a number of valuable insights are given; however, the desire for confidentiality means that some rather wide variation in calculated costs can be found even if one class of compounds is considered. Some general considerations in costing will prove useful, regardless of the process or its novelty.

Manufacturing or operating costs are most often divided into fixed and variable costs. Simply put, fixed costs are those incurred regardless of the volume of production. In the strictest sense, output volume could be zero and only true fixed costs would be expended. Theoretically, fixed or indirect costs must be paid whether or not there is any product output. As such, depreciation, taxes, and interest (on ongoing investment) are examples of such costs. What is immediately apparent is that a host of fixed costs really have a fixed element and another element that is somehow related to output. The somewhat confusing term, semifixed or semivariable cost, is used to detail such costs. One example will suffice. If maintenance supplies are considered a fixed cost (which is reasonable), it is highly unlikely that this fixed cost will be invariant whether output ranges from 0 to 100% of plant capacity. Further, what if plant output is to be 110% of capacity for one quarter of the year? Clearly, cost of maintenance supplies will be approximately fixed for a region from, say, 20 to 80% of capacity. There will be a modest change in slope at 80% and a sharp change in slope at 100%. Therefore, a simple linear relationship is insufficient to describe the true nature of maintenance supplies' cost. The curve might look like that of Figure 12.

Variable costs, on the other hand, are those which are dependent upon production volume. Some clear examples are cost of raw materials, bag or package cost, and cost of operating supplies. Energy is considered a variable cost (which is reasonable), but here, too, one must recognize that there is a distinct fixed element (albeit this is usually a small percentage of the total) and a more significant variable segment. For one to be exact in preparing an estimate, fixed, variable, and semivariable costs should be tabulated for a given and relatively narrow band of production (band referring to a range in percent of total output). When some step-change in output occurs, a new tabulation of costs is made. Then an overall unit cost can be determined more accurately at many different levels of production. Operating costs are not always a linear function of output.

For clarity, we will stay with a simpler two-term concept: fixed and variable. Table 24 shows how costs may be conveniently divided. There are many ways of estimating fixed

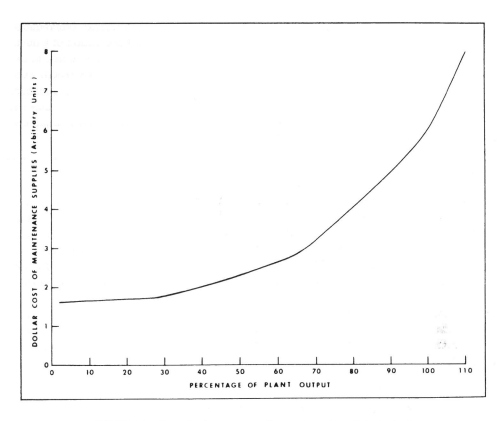

FIGURE 12.   Cost of maintenance supplies vs. percentage of plant output.

costs, which are also referred to as "plant overhead". A relatively simple technique for a new facility is to utilize a percentage (usually 50 to 80%) of total expense for labor, supervision, and maintenance. The most complex method would be to list each item as a line account and budget each based on experience or estimates. Varible costs are determined from pilot plant experience: materials used, yield, loss on processing or packaging, energy use. An example will be given later. Once fixed and variable costs are established, it is possible to determine a "standard unit cost". The individual costs are usually set for a defined operating period; usually this is a quarter or a year. Theoretically, the standard unit cost could be set daily or weekly, but the effort required would negate the needed corollary of a standard unit cost system — an analysis of variance (or displacement) compared to the standard that has been set. The standard unit cost, or standard cost, determines margin levels against a predetermined selling price. For a given time interval having a determined standard cost, operating volume, sales volume, and selling price, it is a matter of simple arithmetic to determine profit levels for that interval.

$$\text{Standard cost} = \frac{\text{fixed cost and variable cost}}{\text{number of units (volume) produced}}$$

Once these determinants are known or assumed, a graphical analysis can be performed to relate fixed and variable costs to profitability (or loss) at various outputs. Breakeven analysis is the term given to this technique. Breakeven analysis is a simple and straightforward procedure. As such, there are limitations to its usefulness, especially in multiproduct plants

Table 24
OPERATING COST DETERMINATION

Fixed		Variable
Salaries	Contracted services	Raw materials (credit salable by-products)
Technical and clerical	Utilities (water, other fixed fees)	Supplies (related to throughput)
Operating labor	Depreciation	Maintenance materials
Engineering	Taxes	Energy (or utilities)
Fringe benefits	Insurance	Packaging
Maintenance costs	Purchasing and receiving	Water
Operating supplies	Postage	Waste treatment (fixed and variable component)
Office supplies	Safety supplies/fire protection	Royalty expense
Rental/leases	Permits and fees	Warehouse supplies
Quality control lab	Advertising (procurement)	Selected equipment rental
Administrative/accounting	Cafeteria	Demurrage
Telephone/telex	Medical expense	Feed product credit
Transportation costs	Technical services/engineering	
Awards	Physicals	
Professional fees	Personnel and training	
	Dues and subscriptions	
	Donations/memberships	
Labor costs include	Payroll burden includes	Energy includes
Training	Vacation, holidays, sick leave,	Electricity
Overtime	company contribution to profit	Fuel (gas, oil, coal, diesel)
Temporary help	sharing as well as mandated	Refrigeration
Time paid not worked	items, pensions, insurance,	Cooling tower supplies
Inventory (time spent)	medical, dental	Utilities chemicals
Labor costs are "fixed"		Utilities' cost normally has a fixed
within a relatively wide		and variable component
operating range		
	Contracted services include	
	Janitorial	
	Guard/security	
	Grounds	
	Rodent/insect control	
	Outside maintenance	
	Trash removal	
	Snow removal	
	Fire extinguishers	
	Laundry/uniforms	
	Computer services	
	Courier	

where assigned overheads can be distributed in different fashions. Still, a relatively simple calculation or graph shows:

1. What volume of product must be made and sold to cover variable and fixed costs for that specific number of units
2. Relationship between variable and fixed costs
3. Impact of volume on profitability (or loss)
4. Impact of selling price on profitability

Assume that certain values are known to a degree of certainty: S = selling price per unit, VC = variable cost per unit, FC = fixed cost per unit, Q = number of units produced

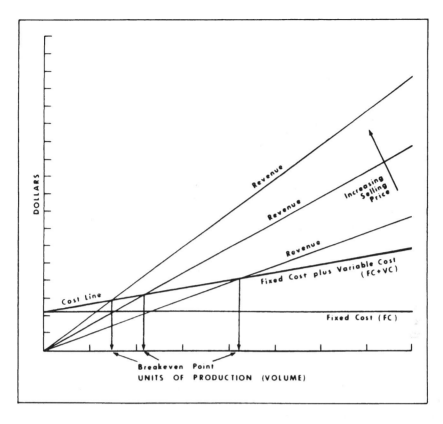

FIGURE 13. Break-even plot.

and sold in a fixed interval, P = profits (pretax) in a fixed interval, VC (Q) = variable cost per interval = VC_I, FC (Q) = fixed cost per interval = FC_I, and Q (FC + VC) = total cost in a fixed interval. Where P = O (the break-even point): SQ = revenue in a fixed interval = Q (FC + VC), or S = FC + VC, Q (S − VC) = FC_I, and VC_I + FC_I = total revenue.

There are various ways of preparing the linear equations; ultimately, a plot is made of dollars as the ordinate (fixed cost plus variable cost as well as revenue as straight lines) vs. volume as the abscissa. A parameter would be selling price. The general shape of the plot would be that of Figure 13.

While the overall standard cost will be a critical element in determining profitability, it should be clear that one standard cost for finished goods would not be helpful in analyzing discrepancies, i.e., actual cost per unit above or below standard. Therefore, a production facility is subdivided into convenient "cost centers", each with its own analysis of fixed and variable costs. Each center has a predetermined output and standard cost; variations can be quickly determined at a specific point in the process and corrective action taken. The major sections of a fermentation plant would be

Fermentation
Extraction
Final purification
Packaging/warehouse
Maintenance
Utilities

One alternative would be to distribute maintenance and utilities costs among the other four plant sections; this would require careful monitoring (meters, flowrates, usages) and equally careful accounting for maintenance costs. The alternative is a monthly conference (or confrontation) to account for over- or underusage. The selected procedure is usually a company policy. At budgetary reviews in the planning cycle, redistribution may occur. It is a tautology, but the closer the costing system and its subdivisions come to reality, the more meaningful the numbers and the responses taken to variations.

Operating costs can be calculated on a unit of production basis or a unit time basis. Whatever is used for cost estimation should conform to the procedures used for recording and controlling costs. Elements of operating costs can best be determined from similar processes run within the company. If no such data are available, many alternatives exist. It is likely that the error in the estimate is inversely proportional to the effort required to prepare the estimate. That is, taking a percentage of inside battery limits cost as total maintenance cost for year 1 of a project is simple, but prone to a potentially large error. A chapter by Black[110] on operating cost estimation is useful for an up-to-date review of techniques. A table and estimate sheet are given which detail individual factors. No table or list can substitute for knowledge of the process and the operation of the planned facility. Every published worksheet must be taken as a guide for individual assessment of operating costs. Table 25 compares cost estimation details and methods of estimation in two references.[109,110]

The total cost of production includes a number of costs which are grouped as "general expense". These are costs which are added to fixed and variable costs, distribution, warehouse and shipping costs, and any contingencies. In the natural products or biotechnology area, the "general expense" is usually very significant. One portion of the general expense is sales and marketing, another part is administrative, and still a third (and major) segment is research and development.

Sales/marketing expenses include salaries, commissions, samples, sales service, travel and entertainment, district office expenses, trade shows, data sheets, and sales aids. Any related outside services (market consultant, segmental analysis, training) are also included. Administrative expense is a prorated portion of corporate overhead which includes salaries of officers, accounting, general engineering, legal services, public relations, financial and computer services, central purchasing, regulatory and environmental affairs, traffic and communications, as well as royalties, rents, and contributions that are corporate and must be allocated.

The research and development costs must be paid in some manner; an allocation to a specific product is reasonable. Very often, a percent of the sales figure is taken and considered the "general R & D expense". Such a procedure can be extended to sales, marketing, and administrative, as well. A corporate history will permit appropriate allocations based upon the historical record. If there is little or no corporate history or if only one product is to be produced, one must make certain assumptions. Overallocation will result in high total cost and may effectively kill every project. On the other hand, underallocation will mean that a certain outlay is not covered by a product cost. The issue is complex, especially in those companies that have a very heavy R & D burden.

Distribution costs are often kept as a separate entity, since the entire load is distributed over many products. Distribution is normally calculated at a specific site; for example, a plant location has one warehousing and distribution expense. Included are wages, salaries, and fringes for this cost center only, maintenance, materials, utilities (this cost center), rental or allocated depreciation, containers, pallets, packaging (if not allocated elsewhere), and freight costs.

A. Raw Materials

An earlier section discusses raw materials, especially natural products. Certain materials

Table 25

COMPARISON OF ESTIMATIONS OF COST OF GOODS

	Black, J. H.	Peters, M. S.
Direct costs		
Raw materials	Price list	ca. 60% of product cost
By-product/scrap	Price list	10—50% of product cost
Utilities	Literature	10—20% of product cost
Labor	Literature	10—20% of product cost
Supervision	10—25% of labor	10—25% of labor
Payroll charges	30—45% of labor and supervision	
Maintenance	2—10% of investment/year	2—10% of fixed capital
Operating supplies	0.5—1.0% of investment/year	10—20% cost maintenance or 0.5—1% of fixed capital
Laboratory	10—20% of labor/year	10—20% of labor
Waste disposal	Lit. or separate estimate	
Royalties	1—5% of sales	0—6% of product cost
Contingencies	1—5% of direct costs	1—5% of product cost
Indirect costs		
Depreciation	5—10% of investment	10% of fixed capital, 2—3% of buildings
Real estate taxes	1—2% of investment	1—4% of fixed capital
Insurance	0.5—1% of investment	0.4—1% of fixed capital
Interest	10—12% of investment	0—10% of total capital
General plant overhead	50—70% of labor, supervision, and maintenance	50—70% of labor, supervision, maintenance, or 5—15% of product cost
Rent	—	8—12% of value of rented land and buildings
Distribution costs		
Packaging	Container costs	
Shipping	1—3% of sales	
General expenses		
Administrative		15% of labor, supervision, maintenance, or 2—6% of product cost
Distribution/selling		2—20% of product cost
Research and development		5% of product cost

may be produced captively at one location and used elsewhere. The transfer price (an intracompany cost) may be either standard cost or selling price or something in-between. At the least, market or selling price should be discounted for sales cost not expended and possibly all sales and administrative costs. If there is no market price, some adjusted figure for an upstream or downstream product having a going market price can be used.

It is important to remember that make-up of certain items must be considered a periodic cost. It is probably best to include these items under raw materials so that all interested parties remember routine consumption (and necessary replacement) of the material. Some examples are

- Catalyst — Catalyst is lost, degraded or inactivated; provision for replacement must be made
- Membranes — Even with washing and regeneration, periodic replacement is needed
- Activated carbon — Some loss occurs even with in-house regeneration

- Ion exchange beads   Mechanical and activity loss occurs
- Adsorbents   Same as above

Although solvents may not be consumed in the process, a certain amount of loss or degradation occurs. This is a cost. Immobilized enzymes or cells will not last indefinitely; periodic replacement is required; this too is a cost.

Transportation costs may be estimated once source and use rate is fairly well defined. Costs can be closely gauged by contacting rail, truck, or barge lines. Shipments from abroad require greater effort and will involve additional costs for customs clearance and port storage; contacting international shippers is a better idea than estimation from the literature. Special handling of hazardous substances may add large amounts to intercountry shipments, since regulations may vary. Regulations require extensive testing of novel intermediates to ensure worker safety. Such testing is lengthy and costly and involves any material which is not transitory within a vessel or reactor. Once again, factors other than rate, yield, and cost must be carefully considered in process design and cost determinations.

B. Utilities

Utilities costs have a fixed and variable element. For plants having a low throughput, the fixed fraction can have a surprisingly high component. Certain utilities have either a connection charge or a minimum monthly charge rate that is independent of usage up to some amount. Even if such a cost were negligible, there is a certain amount of electricity, steam, or water needed to maintain the plant in reasonable stand-by condition for a zero throughput case. (Unless stand-by state is long term, it is not normally feasible to disconnect services and operate from rental units.) The base loading (which involves interior heating, interior and exterior lighting, sanitation requirements, alarms, fire fighting services, communications, and so forth) is the "fixed element". In the usual operating mode, utilities costs, directly related to throughput, are significantly higher. It should also be noted that unit costs for fuel or electricity may drop considerably as usage rises. This has some upper limit beyond which rates are invariant. In some cases, "overuse" or setting new peaks will incur a rate penalty for current or future use.

The problems in predicting future energy rates have been well illustrated during the 1970s and 1980s. Demand projections, even among knowledgeable users and generators of fuel and energy, have been wide of the mark. Pricing projections have been little better. Using inflationary indeces or projections is not a good idea. The best technique is to contact local utilities, review local ordinances, and check published government data for the energy source under study. Rates for industrial utilities are given in Table 26A and B.

Most process plants and all fermentation plants require steam. Low pressure steam is used for steam sterilization of seals, vessels, and piping interconnections (30 to 50 psig); somewhat higher pressures are used in evaporators or in sterilizers (media is heated to 140 to 150°C). Steam generated by boilers may range from 250 to 500 psig. In stepwise reduction, steam may be used for motive power in turbines. Pumping or air compression may be effected with steam turbines. The plant boiler system must be designed for peak load, not average load. Invariably, more than one boiler is installed both for preventative maintenance, unplanned shutdown, and stand-by for peak demand.

Culinary steam is not an absolute requirement in a fermentation plant, unless, of course, a food product is being made. (See Title 21, Code of Federal Regulations 173.310, Boiler Water Additives). However, it may be useful to review the additives used in boiler treatment, since steam does enter the process stream. Even if indirect heat exchange is used for media sterilization, dry sterilization involves introduction of steam into the fermentor.

There is not one cost for steam since a host of variables are involved. However, a typical cost can be developed with clearly defined assumptions. In Table 27, the bases are stated

Table 26A
RATES FOR INDUSTRIAL UTILITIES

	Gulf Coast	East Coast	West Coast
Steam (250 psig), $/M lb	3—6	5—7.50	5—8
Electricity, $/kW hr	0.035—0.04	0.05—0.08	0.04—0.09
Well water, $/M gal	0.40—0.60	0.40—0.80	0.30—0.80
Compressed air, $/MCF			
Process	Use 0.10—0.25		
Instrument	Use 0.20—0.30		
Natural gas, $/MCF	2.50—3.50	3.50—5.00	3—6
Refrigeration (ammonia)	Use $1—2.50/ton/day		

Note: M = 1000.

Table 26B
PURCHASED FUELS MID-1985

	Dollars	Unit
Natural gas		
South Central	2.45—2.70	MCF
North Central	3.50—4.00	MCF
Heating oils — East Coast spot		
No. 2	0.69—0.71	Gal
No. 6, 0.3% S	0.24—0.26	Gal
No. 6	0.21—0.22	Gal
No. 6, 2.2% S	0.20—0.21	Gal

(U.S. Gulf Coast spot price about $0.01—0.02/gal lower)

	$/MMBTU	BTU range
Coal — purchased by utilities		
Sub-bit.		
Nebraska	1.00	8,600—9,000 BTU/lb
Montana	1.55—1.69	
Wyoming	1.65	
Coal		
Pennsylvania	1.12—1.66	12,500—13,500 BTU/lb

(Industrial customers would pay $2—5 more per ton than utilities)

Note: Natural gas has a net heating value of ca. 936 BTU/ft^3 (or 34,866 kJ/m^3). A reasonable value for no. 2 fuel is 131,500 BTU/gal (or 36.6 GJ/m^3) and for no. 6 fuel, 142,000 BTU/gal (or 39.6 GJ/m^3). M = 1000.

and an overall cost is developed in a logical manner. Each plant location and boiler plant design will have a different steam cost. For a very rough estimate that by-passes the detail shown, steam cost (BTU basis) can be taken as two to three times the cost of fuel. If energy cost is $2.50/million BTU, steam cost might be $5 to 7.50/million BTU.

Complete detailing of utilities may be difficult for a grassroots plant, especially if a new product or process is to be introduced. It is not a bad idea to have the required values tabulated, even if all the data are unknown. In that case, gaps in knowledge are, at the least, highlighted.

A picture of utilities requirement would include:

Electricity
Voltage   Emergency generation

Table 27
TYPICAL PLANT STEAM COST

Assumptions

Natural gas usage	200,000 MCF annually
Conversion efficiency	82%
Steam produced	164,000 M lb annually
Natural gas cost	$3.20/MCF
Electricity cost	$0.07/kW hr
5% of maintenance budget to steam generation and distribution	
Fuel oil cost	$0.90/gal

	$/year
Natural gas	640,000
Boiler plant depreciation	50,000
Maintenance (allocated)	30,000
Labor (1 person/shift)	128,000
Treatment chemicals	25,000
Operating supplies	3,000
Supervision (allocated)	10,000
Water/sewer	6,000
Electricity (250 M kW hr)	17,500
Permits, fees, insurance, inspection	11,000
	$920,500

Note: $920,500/164,000 M lb = $5.61/M lb. If 5% of fuel is from fuel oil, a cost increment of $32,300/year is incurred. Cost of steam rises to $5.81/M lb.

Power demand*	
	Power factor
Energy demand*	Specific energy consumption
Unit cost	Transformers
	Capacitors
Steam	
Boilers (number, type)	Demand*
Make-up water	
Pressure	Reducing stations
Fuel	Steam cost
Alternatives	Specific energy consumption
Treatment scheme	Condensate return
Compressed air	
Compressors (number, type)	Demand*
Pressure	Cost
Input filter	
Coarse filter	
Cooling water and process water	
Pumping capacity	Demand*
Wells	Cost
Temperature (year round)	Specific consumption
City water	Treatment scheme
Refrigeration/cooling	
Towers	Demand*

Table 28
OPERATOR REQUIREMENTS FOR FERMENTATION
AND ISOLATION

Workers per shift

Fermentor	0.2—0.3	
Centrifuge	0.1—0.2	
Rotary vacuum filter	0.1—0.2	
Evaporator, ME	ca. 0.1	
Reactor, batch	0.5	(also for media make-up)
Reactor, continuous	ca. 0.2	
Crystallizer	0.1—0.2	
Membrane system	0.2	(also for various adsorptions)
Dryer		
Spray	1.0	(includes some warehousing)
Drum	0.5	
Tray	0.4	
Rotary	0.4	
Boiler room	1	(includes other utilities)

Mechanical   Cost
Capacity   Temperatures (in/out)
Treatment scheme

* Both peak and average.

C. Operating Labor

The best method for determining labor requirements is to list equipment and couple this table with equipment layout, then prepare a manning table. In a fermentation plant, round-the-clock coverage is routine. However, the same number of operators is not required on each shift. A manning table would intersect in some fashion with a batch scheduling sheet (fermentation) and a time schedule for continuous or semicontinuous operations (extraction). It may be possible, for example, to schedule batch charging/sterilization in one or two shifts. The simpler solution of having one person/shift for charging is not cost effective. Furthermore, one extra person on day shift may be most effective for special sampling and testing, vessel inspection, GMP review, and data correlation.

There are published data on "typical" labor requirements for process equipment. These may be somewhat high for an automated fermentation facility. Based upon experience, one can start with manning requirements given in Table 28. A more refined estimate would involve job analysis, scheduling, and a manning table.

In fermentation operations, round-the-clock operation is the norm. Since there are 21 shifts per 7-day week and four operators can cover 20 shifts, there is an assignment of rotating overtime for each person. One every 4 weeks, one operator will cover six shifts. Both rotating and fixed schedules are in use. For the fermentor floor, one fermentor will require at least one person for inoculum development in the plant, media charging, sterilization, and monitoring. The norm, however, is to have six or more fermentors installed. As a first estimate, three operators can be provided. With four operators, perhaps 20 fermentors might be covered. There will be major differences in personnel requirement for a 2-day cycle with special nutrient additions compared to a 10-day cycle with controlled, continuous feed.

In extraction operations, one operator is capable of monitoring more than one unit operation, especially with appropriate controls and alarms. Just how much can be covered depends upon geographic layout, start and stop frequency, need and frequency of cleanouts, and degree of automation.

Supervision is required, but is dependent upon process complexity. If the purification train were relatively simple, one supervisor might be able to handle all operations on a single shift. The usual practice involves a rather complex extraction train which may cover a far larger area than the fermentation plant. In fact, some multiproduct plants have separate buildings for extraction of each of the various products. Therefore, there will be at least one supervisor in each major process segment. In a "simple" train, supervision will cost 10% of labor cost. For a more complex and involved operation, supervision cost may be as high as 30% of labor.

D. Maintenance

Black[110] reports that average maintenance cost in a petroleum refinery are *circa* 5% of investment per year, with a low of *circa* 3% for relatively simple segments, up to *circa* 7% for equipment and processes that are more complex and more subject to mechanical stress or wear and corrosion. The overall range given is 2 to 11% of investment per year, depending upon complexity. For fermentation facilities it would be prudent to use 8 to 10% of investment per year.

There are certain qualitative considerations which should be factored into maintenance planning. Fermentation plants operate on a continuing basis, usually for 11 months/year. Under high demand conditions, no annual shutdown is taken. The need for an excellent preventative maintenance program is obvious. Emergency shutdowns are not only costly to repair, but throughput is reduced. An annual PM program (including scheduled shutdowns) should be established and management should be certain that it is adhered to. There is a tendency to "leave well enough alone" and skip a PM shutdown; this inertial response should be countered. There are many areas of the plant where a redundant or backup piece of equipment can be put into service while a PM is completed; no loss of production occurs and all the pumps or compressors, for example, receive more or less equal attention and equal running time. Should an emergency occur, the stand-by item will have a far greater chance of starting and running at once. The maintenance crew should have selected personnel for routine vibration analysis, routine lubrication, and thermal analysis. The instrumentation group should provide routine calibration of key measuring devices.

The storeroom of spare parts is rather extensive in any continuously operating chemical plant. In a fermentation plant, selection of expensive materials of construction adds to the working capital tied up in this area. Materials' cost for maintenance may make up 40 to 50% of total maintenance cost and spare parts inventory can easily run into many millions of dollars in a large plant. Fortunately, there are computerized systems (from PCs to more complex, multistation networks) to assist in maintenance management. The software available is capable of handling maintenance planning, the PM schedule, tracking of estimated and actual costs, and inventory control and reorder in the storeroom. These systems should be incorporated into planning of a grassroots plant. It is not a great effort to use existing software and utilize a plant-wide computer system if such an approach is taken as an addition to an existing plant.

Cost of a computerized maintenance management program will depend upon plant size, number of maintenance workers and tasks, as well as in-plant capabilities to utilize the software. One company (Patton Consultants in Rochester, N.Y.) offers an expandable program that includes survey, software, hardware, start-up, and training. All elements of a maintenance program are included and price is clearly a function of needs and complexity. In mid-1985, program price range was $6500 to $79,000 with training additional.

Use of computers, plus an overview of modern maintenance techniques, is given in a review article.[112] The requirements and essential elements of a computerized maintenance management system are described. In summary, any such system should provide:

1. Local control whether storage/computation is local or at a distant mainframe
2. Methods of identification/reduction of excessive maintenance costs
3. Forecasting of personnel requirements
4. Early indication of upcoming problems (either equipment or scheduling)
5. Program for reduced downtime
6. Optimization of equipment and tool use
7. Improved planning and scheduling, short and long range
8. Orderly access to equipment and work-order records
9. Stores inventory control
10. Reduction in forms and paperwork
11. Historical data file, repair/cost record
12. Fewer emergency repairs

For a $50 million investment, maintenance cost would be in the range of $3.5 to 4 million/ year. A computer system including software would cost no more than $200,000 and probably less. It is reasonable to expect that annual savings in maintenance cost each year would be equal to, or greater than, that initial cost.

E. Indirect Costs

Fixed or indirect costs must be paid whether or not there is any production output. Depreciation is a major fixed cost. Other such costs are real estate taxes, insurance, and general plant overhead.

Depreciation can be calculated in a number of ways. It is usually taken on a straight-line basis. Where regulations permit, accelerated depreciation can be used for determining taxes. Since legislation determines taxation and also much of the definition of depreciation, the estimate of useful life for cost estimation purposes is very seldom the estimate of life for tax purposes. Depreciation, by dictionary definition, is to lower the price or estimated value of an object. By tax-code definition, depreciation is a process or method by which one establishes a reasonable allowance for wear and tear of an object or property used in a trade or business. The total allowance includes some factor for obsolescence as well as physical wear. By defining depreciation in this manner for taxation purposes, the government is providing a form of tax-free income to the investor. Such an allowance is made to stimulate investment and/or borrowing to that end; hopefully some measure of economic progress will ensue. A depreciable property is carefully defined since the potential for misuse exists. A depreciable property must be used or held for some minimal interval (in the U.S., deter- minable useful life must be greater than 1 year). Further, the depreciable property, which may be a piece of process equipment, a total entity (railcar, truck, car) or a building, must be something that loses value due to inherently natural causes through and during its use. Intangible property is included in the defintion. Examples are patents, copyrights, designs, and even goodwill. It must be recognized that land alone cannot be depreciated. The tax laws in the U.S. now have many classes of depreciable property and expert advice may be needed to catalog items in the appropriate life span. A complex plant put in service after 1981 will show a complex pattern for depreciation and will, in general, involve an accelerated cost recovery calculation as compared to straight-line depreciation. Therefore, in calculating return on investment, the conservative approach would be to use the straight-line method; if the investment is worthy and made, additional economic advantages could accrue by using allowable accelerated recovery methods.

The entire situation of depreciation is extremely complex, since a certain amount of latitude in interpretation is permitted. Acceptable accounting rules and government rules cover a continuum of regulations; these groupings may or may not overlap at all points. Since varying interpretations and methods of application will result in marked changes in reported profits for a given year, there is more than a casual need for the proverbial "good accountant". To create even more complexity, cash flow involves net income, depreciation, and even deferred taxes related to the acceleration method use. Cash flow and reported profits may move in opposite directions depending upon depreciation assumptions and methods used. In a capital-intensive business that has a large segment of valuable intangible property (as patents), it is very difficult to (1) select one standard procedure for depreciation determination and (2) determine how and why certain values were employed on an income statement. This is true for biotechnology companies, as well. It is prudent, before selecting depreciation methods, to consult regulatory or other expertise and establish a procedure that is mutually agreed upon.

Real estate taxes depend upon geographic location, local law, and agreements reached with appropriate authorities. A range of 1 to 4% of fixed capital investment is conceivable with a midpoint value as a good compromise. Selection of a site and discussion with local bodies will fix the percentage.

Insurance rates depend upon technology employed, type of products made, location of the plant, and hazard characterization. This cost is usually about 1% of fixed capital cost.

Royalty and patent costs are normally known for the process involved. If they are unknown, or yet to be negotiated, a figure of 5% of sales price of the product can be assumed. In the chemical process industry, the range is 1 to 5% of sales.

There are other indirect costs that are mostly or totally paid at a rate that is independent of throughput. Office management, control laboratory, special outside services (security, pest control, janitorial service), warehouse, and storage facility maintenance are examples of such costs. A rapid method for preparing a manufacturing cost estimate would lump costs for supervision, indirect payroll expenses, operating supplies, plant office, control laboratory, and general plant overhead into a single figure multiplied by direct labor cost. This lumped overhead would be 70 to 150% of direct labor.

F. Allocation of Overhead Costs

In a single-product plant, no allocation of overhead costs is required. In a multiproduct plant (say, 20 fermentors and some shared and some dedicated extraction equipment), allocation of overheads is necessary. It is also a complex procedure. There is no standardized or defined system of rules for allocation. It must be remembered that no matter how many pages of calculation are presented, there is a certain degree of arbitrariness inherent in the allocation procedure. Any system used must entail some degree of assumption; some independent observer of the plant might readily choose another procedure. With these difficulties in mind, the selected system should make every effort at being rational and should relate, in some more or less obvious manner, to reality. The situation that is to be avoided is one in which a profitable or "popular" product is overburdened in order to protect some other product or class of products. Such allocation distorts information, and sooner or later will have a detrimental effect on the business. In a fermentation plant, overhead allocation can be based upon:

1.   Physical structure or floor area for equipment
2.   Direct labor use
3.   Time equipment is in use (percent of total)
4.   Capital value of equipment in use
5.   Segment function (type/quantity of operations)

If three equally sized extraction buildings are used for three separate products, it would not be too complex a matter to allocate overhead to each "miniplant". Such a circumstance is rare. Still, floor area in use per product can be determined and a percentage figure (of total processing floor area) can be set.

Table 29
PRELIMINARY MANUFACTURING COST ESTIMATES
($ THOUSAND)

Product		Hypothetical		Citric acid
Output (MM lb/year)		1		40
Capital ($MM)		15		57
Raw materials		1,000		4,976
By-product credit		(100)		(600)
Utilities		250		6,040
Labor	8 × $30 M/year	240	(30)	900
Supervision	25% L	60		225
Payroll charges	33% L	80		300
Maintenance	8% I	80		4,560
Supplies	1% I	10		570
Quality control	20% L	48		180
Waste treatment		400		600
		2,068		17,751
Contingency	2% DC	42		355
Depreciation	8 $1/_3$%/year	1,250		4,750
Real estate tax	2.5% I	25		1,425
Insurance	1% I	10		570
Plant overhead	70% L, S, M	266		3,980
		3,661		29,831 ($0.746/lb)

Note: L = labor cost, I = fixed capital investment, DC = direct costs, S = supervision cost,
M = maintenance cost.

Total overhead is allocated by the respective percentage. Hopefully, energy use will be more or less related to floor area; otherwise, metering should be considered for allocation or utilities cost (both direct and indirect). Another method uses time-in-use. Still another employs the capital value of equipment. Allocation will be improved if persons responsible for allocation (and this should include one or more accountants) are made aware of the process. This should not be limited to classroom discussion, but should include a fair amount of walking around and observing the actual operation.

It must be remembered that elimination of a product's manufacture will save the direct costs incurred, but will not save the overhead costs (or indirect costs). It may be that some personnel may be eliminated, although this is not always the case, but service personnel are rarely impacted. Any savings in the indirect cost area will be very small compared to the total dollar overhead burden related to the displaced product. Therefore, this cost must be spread over the remaining products. A thorough analysis may show that a marginal product should not be dropped because of the overhead burden that it carries. The allocation process should occur at least once per year. Manufacturing history can surely be a guide for the process; however, habit is a poor rationale for the important step of overhead allocation.

A typical preliminary manufacturing cost is given in Table 29. Actually, two such estimates are shown to show the effect of volume on overall cost of goods. The hypothetical product is assumed at a throughput rate of 1 million lb/year. (If other items are corrected for appropriate units, the volume figure could just as well be one million doses per year). A 15 million dollar investment is assumed; this is a relatively small plant. The estimated cost of raw materials is $1/lb of product, but even if it were zero, all other costs come to $2.66/ lb of product. Depreciation is about half this ex-raw material total. Raw materials would include cost of any expended material whether a substrate, an adsorbent, catalyst, or de-colorizing agent. One could presume that the finished product, while pure, would require further processing if it were a pharmaceutical. This summary is merely a detailed picture of how unit cost is strongly related to volume. As for our citric acid example, using assumptions detailed earlier, and a capital investment of $57 million plus some expectations of a highly automated process (30 operators), the estimated cost of goods is $0.746/lb. This presents an immediate problem. The selling price of citric acid (early 1986) is approximately $0.83/lb. This level of profitability (see later section on return on investment) may provide inadequate justification for the level of investment required.

VIII. AUXILIARIES

Auxiliaries in a fermentation plant would encompass all buildings, structures, equipment, services, and functions which are not directly involved in the production process. Some examples include power and steam generation and distribution (as well as other utilities), warehousing, waste and water treatment, fire protection, service roads, and communications. Auxiliary building costs are also included. Costs and references have been given above for general items; some detail on selected items is given below.

A. Cogeneration

A fermentation facility is an excellent low temperature heat sink. As such, it is ideally suited to cogeneration. In a commonly used cogeneration topping cycle, fuel is burned in a high pressure fired boiler, and resultant steam drives a conventional turbine generator set which results in electricity (in-house use and sale to utility) and low pressure steam for process use in the plant. Various modifications may be used to achieve the same end. Fuel may be burned directly in a turbine generator with additional heat recovery in steam-generating boilers. Whatever the scheme, overall energy efficiency of 30% for a conventional utility may be increased by 2.5-fold in a properly designed cogeneration system. Utilities must purchase power so generated and a number of systems have been installed in recent years.

The steam user must operate continuously (with little or no downtime), must have a reasonably steady demand, and must be able to utilize low pressure steam. Certain chemical operations fulfill these requirements and fermentation is one such example. Downtime is normally scheduled once per year for maintenance, there is a quantity of rotating equipment which uses electrical or steam drivers, and steam for sterilization and maintenance of sterility seals is needed continuously. Furthermore, average, peak, and low demand does not vary by orders of magnitude.

Cost of a new facility will be higher than a conventional steam-generating plant, but excellent rates of return are obtainable on the differential investment. Availability of alternate fuels offers even greater potential for higher return. A recent article discusses petroleum coke as a fuel in a new facility.[113] The displaced fuel is natural gas. Steam generation rate is 682,000 lb/hr with steam to process of 600,000 lb/hr; net electrical power generation is 19.5 MW. A capital estimate ($\pm 20\%$) is $75 million, but the internal rate of return is 25.6% with the key factor being difference in costs for fuel. The capital estimate includes needed auxiliaries for coke burning and environmental control. Each situation requires a review of investment tax credits, lease potential, depreciation schedule, fuel costs and escalation, electrical utility costs and avoided power costs, environmental and waste disposal considerations, and need for surge capacity for both electricity and steam.

B. Quality Control

Strictly speaking, quality control (QC) is involved directly in production, since both in-process and finished goods determinations are essential for product acceptance. However, it is included here as an added, essential function.

The modern quality control laboratory is not only moving toward more sophisticated measuring devices, but is changing in more fundamental ways. These tendencies have been noted and are continuing:

- More measurements made per man-hour of analysts' time
- Computer networks within labs and between labs
- Computer networks connecting production/packaging with the QC laboratory
- Equipment will incorporate expert systems (pretreatment sequence, experimental design, separation sequencing)
- Extensive use of high volume, data management systems (sample identification and tracking, management, printing, distribution of final results, costing, product tracking, archiving)
- Increasing use of robotics; laboratory operations continue around the clock with or without human intervention
- Robotics with tactile sensing capability and preprogrammed repetitive sequences
- Increasing use of biosensors and highly targeted molecules

There is a well-known relationship between total cost of QC vs. fraction of output inspected. Cost of undetected defects declines as a greater and greater percentage of total output (quantity or units) is inspected. For biologicals, a product defect can be very serious and very costly; however, 100% inspection is extremely rare. Even in that case, some finite number of defects would escape detection. In the usual case, an optimum inspection level, regardless of product characteristics, exists. This may be skewed far to the right (close to 100%), but a minimum cost value does exist. It may be difficult to determine "cost of undetected defects" (it is much simpler to determine "inspection costs"). If one wishes to control cost, however, an effort must be made to find a safe and economic inspection regime.

It is interesting to note that the shape of Figure 14 is rather similar to that developed to determine minimum agitation-aeration cost. The same sort of curve exists for determination of minimum total maintenance cost. If the total cost of preventative maintenance is plotted against degree of preventative maintenance, the result would be an ascending curve (i.e., cost rising as degree of preventive maintenance increases). The cost breakdown and repair would, on the contrary, be represented by a descending curve, falling as preventative maintenance levels rise. There is some point where the sum of repair cost and preventative maintenance cost would be minimum. This would be the economic optimum level of preventive maintenance. The analogy to quality control cost is obvious.

Quality standards are established by one or more of the following: regulatory requirements (governmental bodies), needs of the intended market, competitive standards, sales and marketing demands, dosage form. The earlier the in-process standards and product form and QC standards can be established, the better the process planning and the better the capital cost estimate. This is a truism, but it bears repetition. It is usually extraordinarily difficult and costly to improve product quality significantly once a process is established and running. A process and control system designed to give a purity of 99.5% will require more than a simple addition to give a product with purity of 99.7%. This situation is not universally true, but luck is not the usual accompaniment of process modification.

Responsibility for the manufacture of a bioproduct that satisifes all functional and regulatory requirements resides with the production staff. Quality control personnel may, or may not, report to the production line organization. However, the production organization has final responsibility for meeting specified rate, yield, and quality. (It is probably commonplace for QC personnel to report directly to some other than a direct manufacturing person. In

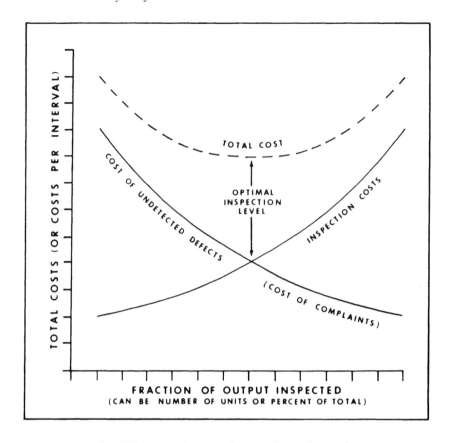

FIGURE 14.   Total cost as a function of output inspected.

this way, on-site pressures for speed and output cannot be exerted.) Quality control personnel should be adept at recognizing and quantifying any deviation from standard at any step in the production process. The sequence of control and observations covers a spectrum of activity:

1. Raw material receipt: include inspection of carrier, product (interior and exterior of container), container
2. In-process control: hazard analysis, critical control points, hold-up (controlled and uncontrolled), calibration of instrumentation, sterility monitoring, exposure monitoring, manufacturing specifications
3. Finished product control: all internal and external specifications, storage and warehouse inspections, stability, monitoring of ambient conditions, timely sampling (stability control), bioactivity, and toxicology
4. Packaging: bags, containers, boxes, final dosage forms, printing (label) requirements for size, content, appearance, inserts, material safety data sheets, weight or dosage control, certificate of analysis
5. Transportation: include inspection of carrier, product (interior and exterior of container or unit load), package itself, outside storage location, product control, recall procedure

Usually, the QC staff is responsible for, or monitors, a number of related activities. Plant sanitation (adherence to GMP and GLP), indoor and outdoor inspections, personnel hygiene (including personnel monitoring), control of insects, rodents, birds (contractor review and control), and guide/contact for regulatory inspections are such related functions. Other critical activities for QC function (either direct or indirect responsibility) include: complaint handling and response, product coding (time of manufacture and expiration date), special release, specific guarantees, product labeling, as well as review of analyses for inclusion in technical data sheets.

There is no such thing as a typical QC laboratory. Layout and equipment selection relates to process and product spectrum. In fermentation operations, rather specialized QC equipment is required. Many commodity products can be analyzed by conventional means (wet chemistry, specific enzyme reactions), but there are many, newer products which require novel and costly apparatus. It may be more efficient to have a centralized location for one or more of these devices (so that a number of lines or plants may be serviced), but a modern, multiproduct fermentation plant might have dedicated instruments for highly specialized assays. Cost of selected instrumentation is given in Table 30. Not every plant will have or need each item. Still, finding more than one such unit at a single location would not be uncommon.

Quality control has benefitted from the growth, or explosion, in software designed for specific applications in that area. These programs incorporate innumerable statistical programs as well as plotting capability in many configurations. Programs are available for personal and minicomputers. Direct and automatic data entry is possible. Optimization routines are available. Various levels of complexity are indicated by range in cost for software only. Prices range from below $100 to over $80,000. A detailed tabulation and description of available programs is given in *Quality Progress* of March 1985 (entitled "The ASQC Software Directory" beginning on page 27).

A number of laboratory information management systems have been devised. The Beckman Computer Automated Laboratory System (CALS) is described in great detail in an article.[114] Whatever system is ultimately employed, the capabilities given in the article should exist in the hardware and software packages:

- On-line, real-time data acquisition from almost any digital or analog electronic instrument
- Chromatographic analysis by laboratory-specified methods
- USP system suitability analysis with documentation of column conditions
- On-line data base maintenance, retrieval, and reporting for sample tracking
- Invoice generation
- Year-end analysis with trend plots
- Instrument calibration and PM scheduling
- Content uniformity analysis
- Drug stability monitoring
- Statistical analysis
- Table/capsule dissolution studies
- GMP and GLP compliance features — automatic date and time stamping, archiving/recall, nonobliteration of data/security system, audit trail, and configurable verification, validation, and approval

(See also section on "Computer Control" for regulatory approach to these systems). Examples are given for each of the above with some representative tables and plots shown.

C. Waste Treatment

The entire subject of waste treatment can occupy many volumes. There is no intent to cover the many aspects of a complex technology. Some consideration must be paid to this auxiliary operation, since associated costs are so high for the usual fermentation production plant. If a treatment plant must be built as part of a grassroots facility or an expansion, investment for this one section can be an appreciable percentage of the total fixed capital.

Table 30

COST OF SELECTED ANALYTICAL INSTRUMENTATION

			Approximate cost ($)
Baird	Air plasma spectrometer model APS-100	Any 6 (of 65) elemental constituents	70,000
Spectro	Spectroflame-ICP	Elemental analysis (64 elements with 10-element base capability)	40,000 + 1,000 per additional element
Perkin-Elmer	ICP/6500XP plasma emission spectrometer	20 elements/min	80,000
Perkin-Elmer	Model 5100AA	Flame, graphite furnace, or Zeeman	39—60,000
Grun Anal. GmbH	Model SM20 Zeeman effect	Determinations on solid samples	55,000
Shimadzu	Model AA-670	Optimal conditions for 61 elements stored in memory	20,000
Automatik Machinery	IROS-100 FTIR	On-line control capability	65,000
Bio-Rad	FTS 40 (FTIR Spectrometer)		39,500
IBM	Model IR/44		36,300
Perkin Elmer	IsoPure LC System	Separate/purify complex, labile biomolecules	18,000
Dionex	Series 4000i ion chromatograph	Separate organic/inorganic ions simultaneously	16,200
Applied Biosystems	Model 130A Separation System	Isolation/purification of proteins and peptides	37,000
Supercritical fluid chromatography (SFC)			34—40,000
Analytical GC			6—14,000

Adapted from *Chem. Eng. News*, March 24, 1986.

Operating cost will invariably be a factor; even if permanent operators are not required (this would be the exception), there will be continuing need for sampling, analytical support, and maintenance. If effluent is transferred to a local or regional authority, investment is avoided but operating costs will be high. Whatever the circumstance, treatment economics are significant and cost reduction should occupy planners as well as persons involved in plant construction and operation.

In general, fermentation processes produce large amounts of aqueous waste with appreciable BOD or COD loadings. Solids are often present. Moreover, and again in general terms, the waste streams resemble municipal wastes in treatment needs and options. Cost of the building treatment facilities coupled with prohibitive user fees have led to most fermentation plants having pretreatment or total treatment plants on-site. Plant effluents undergo settling, aerobic and/or anaerobic digestion, filtration, oxidation in ponds, specific tertiary treatments, or any of a combination of these operations.

The fermentation step probably produces the greatest treatment load in that alternate metabolites are produced, unassimilated substrate is present, and many cations and anions are present unrelated to the product. While many vitamins, cofactors, precursors, and cations are essential for product synthesis, the excess of these materials must be disposed of in some manner. If cells themselves are not the desired product, the cell mass becomes a major contributor to BOD and suspended solids (SS). One last point concerns the fermentation step. A major upset in pH control, temperature control, agitator, or aeration may mean that productivity is nil. One is left with a large volume of a carbohydrate-salts mixture. Should the batch become contaminated and product titer is too low for economic processing, there is another case of major loss. Not only is productivity lost, but large treatment costs must be incurred with no concurrent product recovery. Rarely is a contaminated batch resterilized for reinoculation; most often the entire batch ends at the effluent treatment system.

The effluents generated by the extraction steps will vary with the process employed and generalities are not very useful. It should be recalled that regenerants must be disposed of, that solvents are not totally recovered, adsorbents and carbon degrade physically with some portion ending in effluent, that processing aids must be disposed of, and that as a rule, every item charged to a process ends up in an air stream, a product stream, or an effluent stream.

Even if one or more process streams can be modified, treated, or simply shipped as an agricultural commodity (soil treatment, fertilizer, animal feed), a fermentation plant will generate appreciable effluent that requires some treatment option. Sanitary waste is produced by plant personnel, wash and shower facilities, cafeterias, and outlying structures (security, gate houses, weigh stations). Sanitary or domestic waste can be handled in a separate system, sent to a local authority, or mixed with plant effluent for treatment. Storm sewers are used to minimize volume impact on treatment plants. Design must assure that no process streams or spills enter the storm drain systems. Further, diking should provide for containment of materials stored in above-ground tanks. Process wastewater will contain foreruns or tailings in many unit operations, wash streams, many regenerant streams, leaks, contaminated condensate, some discard streams, and spills. Some washout is required after each batch; unless this dilute stream is sent to processing, it will become part of the plant effluent. Washout refers to vessels, mix tanks, and transfer piping. Cooling tower discharge or cooling water discharge cannot normally be sent directly to a waterway. Blow-down volume contains various treatment chemicals. Even if all heat exchanges are indirect, leaks are possible which will contaminate even once-through cooling water.

Another contributor to the treatment load is the sum total of all cleaning operations. Not only are chemicals, detergents, surfactants, and disinfectants included in this grouping, but one must include water or condensate that is contaminated with a process stream or debris.

One method (noted earlier) that deserves early consideration is the "creation" of salable by-products from the process. Certain streams that require disposal may be quickly and inexpensively converted to a useful animal feed product or a crude fertilizer. Concentration or simple separation, with or without pH adjustment, may be sufficient to give a useful product. Important criteria are stability of the product (resistant to chemical or microbiological degradation) and cost of transportation to the user. Alternate use consideration should be applied to every large volume or high load stream at any point in the process. With treatment costs rising and with ever more stringent environmental controls, recycle and reuse in another industry is a preferred choice. Treatment of an effluent stream to prepare the product must be considered merely another economic constraint and cannot be dismissed because it adds to the cost. The future alternative may be less pleasant. In the same vein, reuse of water or condensate which is not too badly contaminated (in the broader sense) must be considered. Again, an economic balance must be drawn; however, water is neither "cheap" nor abundantly plentiful and many municipalities have a flow factor attached to treatment costs. It costs money to send pure water to many treatment facilities.

Fermentation plant wastewater will contain appreciable contaminants in the form of (most of items below):

1. Soluble organics which support excellent microbial growth. This gives rise to BOD (biological oxygen demand) which is a measure of potential for oxygen depletion in a waterway. One pays for "pounds of BOD" on some unit time basis; limits are placed on concentration or mass quantity.
2. Soluble or insoluble compounds which react in an acid dichromate test. This gives rise to COD (chemical oxygen demand). COD includes both a biodegradable and a nonbiodegradable element.
3. Suspended solids (SS). Suspended solids may be cells, cell fragments, precipitates due to stream mixing, filteraid, carbon, or adsorbents. There is usually payment required for "pounds of SS" per unit time and a maximum exists for concentration or quantity per unit time.
4. Inorganics which arise either in fermentation or extraction. Phosphorus (as phosphate) is a common contaminant that is receiving increasing attention by regulators.
5. Ammonia (NH_4^+) which came from medium charge or during processing (example is use of $[NH_4]_2 SO_4$ or stripping into condensate). More and more authorities are limiting ammonia in effluent to a treatment plant; even so, there is a charge based on content.

There are also limits on pH and temperature and there may be a dissolved oxygen limit (must be above some minimum) on treated effluent. There is usually a charge for volume that may not be linear, rising geometrically as effluent volume rises. There is normally a fixed element in the charge related to capital recovery (assume off-site treatment) or merely to insure some payment regardless of flow. One can generalize and state that regulatory bodies can and do relate payment to concentration and/or mass of constituent per unit time along with requested adherence to ever-more stringent limits on one or all key entities in the discharge stream.

Pretreatment steps in common use include:

1. Neutralization: preferably various streams can be blended to give a pH in the acceptable range.
2. Equalization: a large holding tank or basin is used to stabilize flow and concentrations. Surge capacity allows for leveling or averaging of content, shocks on subsequent steps are minimized.
3. Skimming: removal of oil, grease, defoamer, or any separate liquid phase that floats on the water

Table 31
ULTIMATE WASTE DISPOSAL METHODS

	Evaporation ponds	Injection into wells	Landfill	Agricultural	Ocean
Liquid	Likely	Problematic	Possible	Possible	Costly
Sludge	No	No	Possible	Problematic	Costly[a]
Combusted ash	No	No	Yes	Problematic	Costly
Cell cake or paste	No	No	Possible	Possible	Costly

[a] Sludge can be combusted on site or nearby or can be burned on a vessel offshore. Incinerators commonly used for sludge combustion are multiple-hearth and fluid bed types.

Primary treatment can include:

1. Grit removal
2. Flocculation
3. Sedimentation
4. Chemical addition (as phosphorus removal)
5. Preaeration

Secondary treatments include:

1. Aerobic digestion (lagoons or tanks)
2. Anaerobic digestion
3. Trickling filters
4. Sludge treatments (lime, alum, iron salts, heat)

Since fermentation plant effluents normally contain high levels of biologically compatible compounds, aerobic digestion is a prime option.

Final treatments would be required for special circumstances and would include:

1. Ozonation
2. Chlorination
3. Combustion
4. Carbon treatment
5. Ion exchange

Additional processing steps for fermentation effluents might include ammonia removal (steam or air stripping), denitrification or specialized treatments for removal of phosphorus, dialysis or reverse osmosis, or evaporation. These steps could occur anywhere from pretreatment step to a unit operation that occurs on treated effluent. The processing step and its location in the sequence would depend upon nature of impurity removal, volume of stream to be treated, and economics.

Ultimately, even if a complex treatment scheme is employed, some residue remains. The residue can be considered that product form that resists further treatment and that must be disposed of in some convenient manner. On-site burial was once a permitted option; requirements and restrictions have changed that situation dramatically. Of course, no clear-cut answer to disposal can be given since volume and composition of matter are critical determinants. In general, however, some options are more reasonable than others (see Table 31). Many fermentation plants that have a resulting sludge (aerobic digestion or other processing) are equipped with a dehydrator and/or furnace to burn the solid residue. Even though volume is sharply reduced, there remains the need to dispose of the ash. Injection of liquid streams is becoming less and less of a viable option due to environmental impact. Landfill requirements are becoming very stringent, since leakage from older landfills has been a serious problem. Ocean discharge is not only very expensive, but environmental groups are working to move such dumping further and further from shore. Agricultural uses of various partly or fully treated wastes offer an attractive option for both generators and farmers. Cost would be attractive and, in many cases, valuable nutrients are added to the soil. Compatability exists. However, certain constituents (bioactive compounds, heavy metals) must be monitored very closely.

A paper summarizes various approaches taken to the wastewater problem.[115] Effluents are a significant problem and designers of the process as well as manufacturing personnel must be concerned with investment and operating costs. The authors state that cost of a waste treatment facility on-site of a fermentation plant could range from 10 to 20% of the total capital investment. Since environmental regulations play a key role in defining technologically sound and economically viable options for reuse or treatment, these regulatory needs must be incorporated into each stage of design. Development of alternate use should proceed in parallel.

A number of different options were reviewed by these same authors. Surprisingly, investment varied by less than a factor of two for a given hypothetical plant (the case chosen was for alcohol production):

Option	Investment	Operating cost
Conventional aerobic	1.00	0.50
Existing system upgrade	1.66	0.72
Conventional anaerobic	1.55	0.64
Condensate Treatment		
Electrodialysis	0.5	0.81
Ion exchange	1.10	0.50
UV plus peroxide treatment	1.75	0.70

Units are given on a comparative basis (dimensionless). Comparison unit for capital cost is different from that for operating cost. Included are facilities for sale of condensed by-product.

There are many options open for potential disposal routes for wastes generated in a fermentation plant. If the presumption is that there is no hazardous, toxic, or regulated substance in the discharge stream, all options are possible, even direct agricultural use. Constraints enter as required treatments increase. The potential routes for disposal are shown in Figure 15. Presumption is that physical-chemical treatment of effluents will be insufficient; however, even if simple treatments were feasible, the pathways can be readily modified. The goal is avoidance of tertiary or special treatments.

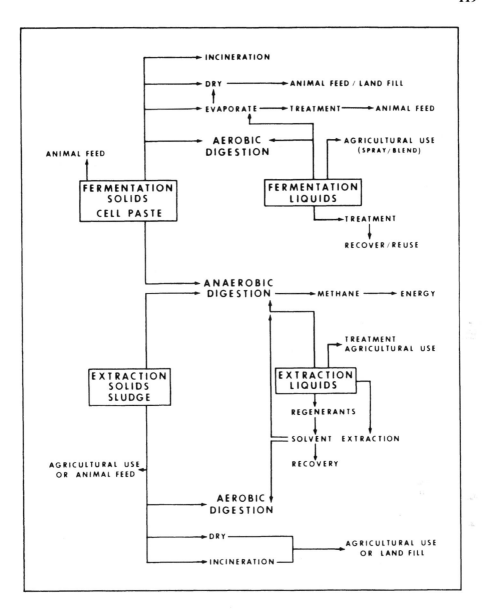

FIGURE 15.   Fermentation plant effluent streams — potential routes of disposal.

Chapter 3

REVIEW OF PUBLISHED ECONOMIC DATA

The fermentation industry may be divided into these groups.

Start-up companies — In the last decade or so, many biotechnology firms have been funded to achieve commercial applicability of what is loosely termed genetic engineering. A few compounds or classes of compounds (highly selective goal orientation) are under intensive development; production facilities are limited.

Biomass production and solvents — Certain firms are interested in production of yeast, yeast extracts, or single cell protein (SCP) for animal feed or human use. Substrates run the gamut from sugar, molasses, sulfite liquors, ethanol, methanol, and, more long range, methane and agricultural wastes. The cellular material itself is often the product, but solvents are also of interest.

Pharmaceutical and fine chemical companies — These firms are large, highly integrated, and have major fermentation capacities (of the order of 1 million gal or more each). Products are organic acids, amino acids, antibiotics, vitamins, vaccines, and other therapeutic agents. Research costs are very high and many screens and searches are underway in numerous areas of physiological response.

Food fermentation — This grouping encompasses beer, wine, pickling, vinegar, alcohol, select dairy processes, and preparation of traditional fermented foods. Economic analysis follows that of the food industry.

It is interesting, even if obvious, that the amount and quality of economic data and analysis are very different for these four groups. In the first group, hardly any data are published. Information concerns funding of the company, allocation of resources for tax purposes, R & D partnerships, and invested cash. Only recently have the more mundane requirements for production capacity, sales and marketing, and regulatory affairs been noted.

It is the second group that has the most economic information in the open literature. This may be due to a number of factors. First, many governments have funded work in this area at universities and laboratories. The results are very often published. Second, substrate costs (which are usually a major part of total cost) are commonly known. Third, the process is relatively simple and competitive pricing is well known. Lastly, there was a general, if erroneous, opinion that microbial biomass would "feed the world." While this idea stimulated publication and may yet come to pass, the commercial success has been slower in coming. Releasing economic data carries a low risk.

The fine chemicals business fosters secrecy and members seek competitive advantage at each step of commercialization. Economic analysis comes many years after novelty has worn off and most member companies have come up with equal or comparable technology. Some rare examples of cost analysis do exist and will be discussed below.

In the last group, investment and economic decisions follow competitive businesses whether or not they are fermentation based. Further, government subsidies and taxation policies (relative to agricultural commodities and consumed beverages, respectively) render financial analysis beyond the scope of this discussion.

While we are interested in economic analysis in group 3 products, there may be some useful information in a review of articles related to group 2 processes. One such extensive analysis was published in 1982 by Chem Systems, Inc.[116] Since the study was done for the Department of Energy, it is available to the public. The study was performed to assess the technology of producing mixed solvents (mainly butanol/acetone) from renewable agricultural sources. The design basis covers pretreatment of a wood substrate, enzyme hydrolysis, fermentation, and purification.

Assumptions in the study were

1. Feedstock abundantly and inexpensively available in U.S.
2. Acid prehydrolysis, enzyme production, enzyme hydrolysis of substrate are all required
3. Rate and yield data used attained at various laboratories
4. Initial sugar concentration either 3 or 5% (two cases given)
5. Residence times 72 hr (3% sugar) and 48 hr (5% sugar)
6. Yields: 27.5% (5% sugar) and 32% (3% sugar) — batch cycle
7. Purification methods follow conventional technology

A process description, including flow diagrams, is given. Operating parameters are listed. Design basis is 50 million gallons per year of mixed solvents at a plant site on the Gulf Coast of the U.S. (mid-1982). A cost of production estimate is given for each case; included are

1. Raw materials cost with unit prices
2. Utilities cost (power, cooling water, process water, high and low pressure steam) with unit prices
3. Operating costs (labor, supervision, maintenance)
4. Overhead expenses (factored)
5. By-product credits (CO_2 and SCP)
6. Sales price at 10% DCF (discontinued cash flow)

The overall results show:

Case		Low yield	Maximum yield
Total fixed investment ($MM)		190.1	206.5
Total cost of production,	¢/gal	220.9	237.4
Credit	¢/gal	(23.2)	(22.3)
Net cost of production,	¢/gal	197.7	215.1
Selling price (10% DCF)		258.0	282.6
Conventional route to solvents		Acetone	Butanol
Selling price (10% DCF)		232.4	235.1

It was also deemed important to separate plant sections, for the fermentation study, and determine percentages for each segment. The analysis pointed to the purification step as the most significant section relative to total cost. Energy consumption was significant in this section. A research direction is indicated. A short analysis of the conventional route is given. The latter holds an advantage of about 10% in overall cost. Suggestions are made which would allow parametric analysis in order to establish and rank R & D options. Seventeen references are given. The factors used in overall economic evaluation in the Chem Systems report are compared to two other references (discussed below) in Table 32.

An extended discussion is given here, because the report is a good one and embodies many features that should not be overlooked in any economic analysis. The facts that the processes were actually run (through the first half of this century using different substrates) and that many references are available allow a more reasoned approach; fewer assumptions are required. However, the report gives an essential step-by-step procedure which should be followed in preliminary analysis; refinements can be added as evidence becomes available. Interestingly enough, a paper on the same subject[117] that was published 2 years earlier was not included in the bibliography of the Chem Systems report. Lenz and Moreira prepared

Table 32
FACTORS UTILIZED IN ECONOMIC EVALUATION

Ref: Moo-Young et al.	Chem Systems, Inc.	Keim
Capital equipment cost	Capital cost	Investment
Electrical energy	Power	Power
Natural gas	Cooling water	Chemicals and miscellaneous
Pulpmill sludge (RM)	Process water	
Fertilizer (RM)	Steam (high/low pressure)	Steam
Process/cooling water	Raw materials	Raw materials
Weight loss		
Protein in product		
Selling price	Selling price	Selling price
Depreciation	Depreciation	Depreciation
Direct labor	Labor ($26M)	Payroll
FCI/CEC	Foremen ($29.6M)	
Credits	By-product credits	By-product credit
Start up cost		
DCFR basis	DCF basis (10%)	"Hurdle rate" (40%) (preoverhead, interest, tax)
Scale factor		
Fringe benefits		Scale factor
Supervision, QC (20% of direct labor)		
Maintenance (5% FCI)	Maintenance (6% ISBL)	Maintenance (4% investment)
Insurance, taxes (2.5% FCI)	Direct OH (45% labor/super)	Miscellaneous (1% investment)
Supplies (10% maintenance)	Plant OH (65% operating costs)	
General expense (5% sales)	Insurance, tax (1.5% FCI)	Tax, insurance (2% investment)
Income tax (44% net income)		

capital and operating cost data for a plant that would produce 45×10^6 kg/year of solvents. Using a mixed solvent density of 0.801, the volumetric output is 14.8×10^6 gal/year or 3.38 times smaller than the plant described earlier. The "new total capital investment" in the Lenz and Moreira paper is given as $26.3 million (whey feed) including land, which is significantly lower than the figure estimated for the larger plant, even correcting for size. The difference in costs can be explained partially by the fact that the Chem System plan calls for prehydrolysis of a cellulose substrate. Some 40% of the inside battery limits cost relates to enzyme formation and hydrolysis. Lenz and Moreira utilize molasses or whey. A few years' difference in the basic cost estimate could result in appreciable inflation as well. With these points in mind, it is still of interest to compare the analyses. The metabolic pathway is given as well as a complete energy analysis of the fermentation. The final solvent ratio is not very different in the two cases, but some discrepancy in the percent ethanol exists:

	Chem Systems (%)		Lenz and Moreira (%)
	Max yield	Min yield	
Butanol	72	61.7	60
Acetone	25	31.8	30
Ethanol	3	6.5	10

The base case involves:

1. Traditional molasses feedstock
2. 45×10^6 kg/year of ca. 15×10^6 gal/year solvents

3.   Location in southeastern U.S.
4.   300-day/year operation

There are 4 cookers (sterilizers) and 16 fermentors (1514 m³ or 400,000 gal each). Cycle is 48 hr. Hydrogen, CO_2, and residual broth solids are recovered for sale. The relative magnitude of equipment costs is as follows:

Operation	% of total cost
Feed preparation	10.7
Fermentation	39.1
Distillation	14.0
Product storage	7.4
Stillage concentration	28.8
	100.0

After finding that the base case gave doubtful economics, whey was inserted as an alternate substrate. Even with added capital for ultrafiltration and waste treatment, positive return was generated. The net return on investment is 30.8%. The authors do mention the fact that lignocellulosic residues may be particularly attractive for this fermentation. Capital costs are not mentioned. The base case equipment cost (early 1979 values) is given as $5.139 million; this is given as delivered and installed with added costs for instrumentation, electrical, piping, etc. Of the equipment cost, each fermentor is costed at $125,000 and each multieffect evaporator is $212,000. Further, net production cost is about $1.40/gal vs. approximately $2.00/gal in the Chem Systems report. While no definitive answer as to correctness is implied, there is no industrial-scale acetone-butanol fermentation plant in operation in the U.S. As was the case for other fermentation products displaced by chemical synthesis (as riboflavin), long-range prediction is fraught with danger.

A more recent paper describes an industrial research effort.[118] A novel strain was isolated that exhibited higher solvent tolerance, increased stability, and higher fermentation rate. Accordingly, a multistage continuous fermentation process was devised. Corn endosperm which is partly acid hydrolyzed (to *circa* 10 D.E. syrup) is the substrate. Cornsteep liquor is an additive. Butanol, acetone, ethanol, acetic acid, and butyric acid concentrations in the final whole broth stream are 1.5, 0.6, 0.2, 0.18, and 0.08 wt% respectively. Broth is processed in a broth still; distillate is sent to an ABE still where a split is made between ethanol and water. Distillate from the ABE still goes to an acetone-ethanol still for separation. Bottoms plus distillates from butanol-water stills enter a decanter for phase separation. Insoluble protein, fiber, and cells are flocculated, dewatered, dried, and sold as a feed by-product. Solvent recovery processes have been designed and optimized by computer simulation; other parts of the process have been demonstrated.

The economic evaluation is based upon a 200 million-lb/year grass roots *n*-butanol plant. Capital investment was estimated at $119.3 million (1984 basis, error of −15 to +30%). Inside battery limits are subdivided as follows and the other elements are included:

	$ Millions	% of total
Feed preparation	15.06	
Fermentation	8.97	
Protein/cell recovery	9.53	
Distillation/solvent recovery	23.80	

Control room	3.00	
	60.36	50.6
Outside battery limits	46.28	38.8
Start-up expense	2.52	2.1
Working capital	10.10	8.5
Total investment	119.26	100.0

Manufacturing costs are based on corn at $3/bushel. Total variable costs are $0.082/lb butanol, total fixed costs are $0.042/lb butanol, capital charges (30% of total capital) are $0.179/lb, giving a rational price of $0.303/lb of butanol. Sales and general and administrative expenses are not included. Economics and assumptions are detailed as are comparative economics for a petrochemical process (rhodium catalyst) to give the same output. Total capital investment for the petrochemical process is far less than the fermentation process; the estimate of $69.8 million is 58.5% of the biochemical process plant. The rational price (cash costs plus capital charges) is $0.290/lb butanol with propylene at $0.19/lb, accounting for 45% of the rational price. Sensitivity analyses of the processes with various assumptions are plotted. Even with potential errors in assumptions, there is a large region of overlap in the rational prices by either process. Directions for future effort are discussed. It is noted that implementation of the fermentation technology is dependent, not only on technical progress and raw material costs, but also growth in future demand for these solvents. It is reasonable to predict that large-scale fermentation synthesis for these solvents will not be a near-term (5-year) phenomenon.

Another series of papers[119-121] originating from the University of Waterloo in 1979 and 1980 can also serve as a guide for economic analysis. The process itself is relatively simple and falls into the category of biomass production from agricultural residues. A cellulolytic fungus, *Chaetomium cellulolyticum,* grows (sterile process) on raw materials which are widely available in large quantities and are generally considered by-products. A relatively high protein, highly digestible product results which can replace soybean meal and/or yeast in feed rations. The process involves (see Figure 16):

1. Thermal and/or chemical pretreatment
2. Aerobic fermentation (nutrient supplementation)
3. Separation of solids from broth and drying

The final product underwent a number of animal tests and was deemed suitable from viewpoints of palatability, safety, digestibility, and nutritive value. The process cost is said to be low and return on investment (ROI) calculations and discounted cash flow return (DCFR) after taxes were performed utilizing various computer simulations.

Bioconversion of corn stover, straw, manure into SCP

NPK source	Product	DCFR (%)	ROI (%)	$ Can (thousands)	
				Capital	Operating/year
Fertilizer	73% DM·SCP 3.6 tons/day	30	41	465.5	332
Manure (digested)	60% DM·SCP 3.5 tons/day	30	41	469.3	265.7

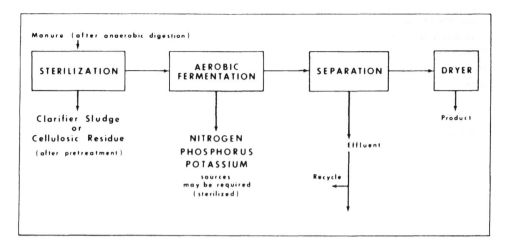

FIGURE 16.   Schematic flow of the Waterloo SCP process.

Manure (digested)	Undried, unfiltered 60% DM·SCP 3.0 tons/day	30	41	378.5	232

If a 20% DCFR (or 29% ROI) were acceptable, then allowable output rate falls by about one third and capital costs fall into the Can $313 to 318,000 range. The economic analyses above were performed in April 1979 and are based upon Canadian figures for the selling price of spent brewers' yeast. Amortization is 10 years. In the next paper noted,[120] cellulosic wastes from pulp and papermill operations are used as substrates. With the assumptions made, the process was deemed "quite profitable". Assumptions and conditions are listed:

Clarifier sludge, 10 tons/day
Capital equipment cost (CEC) from quotations and indices
Scale factor (scale up or down), 0.6 for CEC
Total fixed capital (FCI) = CEC × factor; factor usually 3 (2 and 4 tested)
Product, 7 tons/day containing 25% protein (56% SCP biomass)
Feed solids, 3%
Air (fermentor), 0.2 vol/vol/min
C/N in feed, 8

The following plots were generated to exhibit sensitivities:

1.   DCFR (%) vs. capacity in tons/day; additional parameter is selling price of 45% protein material
2.   DCFR (%) vs. FCI/CEC ratio (from 2 to 4); additional parameter is selling price of 45% protein material
3.   DCFR (%) vs. sludge disposal credit with various outputs as a parameter
4.   DCFR (%) vs. wages with various outputs as a parameter
5.   DCFR (%) vs. operators/shift with various outputs as a parameter

Results show that the process is attractive if selling price is high enough. Then the (1979) selling price of soybean meal was attractive. The fixed investment required was estimated at Can $563,500.

The last of the three papers carries out extensive sensitivity analyses. A detailed list of items and assumptions is given to complete sensitivity analysis on a 5-ton/day corn stover treatment plant. The net product value (NPV which should not be confused with net present value) is calculated as "product selling price less crop residue cost less nitrogen-phosphorus-potassium cost." NPV is plotted against many variables with percent DCFR as a parameter.

A number of scenarios were attractive *if* the product was competitive with soybean meal (cost at that time Can $340/ton at 45% protein). The outline and methodology are worth consulting for general application in performing economic calculations. It must be noted, however, that in mid-1985 there were no operating licensees using the aforementioned technology.[122] In fact, a redirection of the marketing effort was underway. Rather than continue to attempt to license in Europe, Canada, or the U.S., a new thrust toward the Third World would be made.

Potentials for various substrates in SCP production are reviewed and detailed in an article by Abbott and Clamen.[123] The evaluation of substrates was related to costs for the substrate and associated costs of required oxygen and cooling. Costs in the paper relate to operating costs only and a fixed cell production rate was assumed. Costs for oxygen transfer and heat removal were based upon rates at that time. Using various published correlations, plus reasonable assumptions and yield values, it appears that glucose is the most economical substrate. Methane, ethanol, and isopropanol are undesirable substrates from an operating cost viewpoint. However, overall economy of operation must include long-term price and availability, investment, and postharvest treatments as noted in the article. Still, when alternatives are available, the steps proposed can be used as part of a screen.

Economics of SCP production are also reviewed by Moo-Young.[124] Costs of conventional and unconventional proteins are listed and estimated costs are given (ratios or percentages) for different substrates used to produce yeast. Substrates listed are paraffins, alcohols, gas oil, molasses, and acetate. A comparison of various microorganisms and substrates shows a relatively wide variation in cost distribution. Raw material cost remains the most significant fraction of the total — 44% of total cost for bacteria acting on bagasse, up to 77% of total cost for yeast grown on ethanol. Utilities cost shows just the reverse with 12% of the total for ethanol-grown yeast up to 37% of total for the bacteria-bagasse combination. Percentage of total investment is estimated for various scenarios. Fermentation equipment, as one might expect, requires the largest portion of total capital. In the case of bagasse, only 30% of investment cost relates to fermentation. For four other combinations — yeast/paraffin, bacteria/methanol, yeast/ethanol, or fungi/waste streams — fermentation investment cost is 43 to 51% of the total. Comparative utility costs are given; again, fermentation cost is the highest of all utility factors, accounting for 54 to 77% of the total. Relative costs are also derived for individual items of equipment in each section of the plant.

Major efforts, including massive investment, have gone into production of fermentation ethanol around the globe. An extensive economic analysis is given by Fong et al.[125] All substrate-process combinations were designed to result in 25 million gal/year of 95% ethanol. Sugar crops (including sorghum), crop residues (wheat straw, corn stover), and aquatic plants were the starting materials for C_6 sugar production. Both batch and continuous fermentation processes were analyzed. The total plant consisted of a sugar production facility which delivers product "over the fence" to an ethanol plant. The latter encompasses ethanol fermentation, ethanol recovery, utilities, and general service facilities (plus sugar concentration if needed).

Facilities Investment Cost (Millions of dollars)

	Sugar production	Batch fermentation	Continuous fermentation
Sugar cane, conventional	49	23	11
Cane/sweet sorghum	51	23	10
Wheat straw hydrolysis	58	27	14
Stover hydrolysis	50	21	10
Aquatic plant hydrolysis	28	32	

The relationship between plant facilities investment (PFI) and total capital investment (TCI) is given as: TCI = 1.3 (PFI). Figures in the table above represent TCI. All costs are in first quarter 1978 U.S. dollars. In the "regulated venture case", base case costs for ethanol (FOB plant) ranged from $1.85 to 2.46/gal. In the optimistic case (best sugar cost and continuous fermentation), ethanol cost was $1.10 to 1.55/gal. In the *non*regulated venture situation, range for base cases was $2.20 to 2.95/gal with the optimistic case giving $1.30 to 1.96/gal. At the time the estimates were prepared, ethanol selling price (95%) was $1.20/gal. Since the work was funded by federal fuel and energy offices, detailed background calculations are probably available for interested persons. The unstated conclusion of the paper is that considerable process improvement and cost reduction will be needed before any sort of transition to fermentation ethanol occurs.

Keim[126] provides a thoughtful review of fermentation ethanol production to 1983. Capacities of major U.S. installations are given. The total annual capacity (not necessarily output) is given as 445 million gal/year in mid-1983. The review places alcohol in perspective relative to fuel requirements, scale factors, and cost effectiveness. The author also notes that if all the corn harvested annually in the U.S. were converted to alcohol, that quantity would provide less than one quarter of national gasoline needs. Raw materials and special preparatory requirements are given. Various important process parameters and process improvement concepts are given. The key elements in determining underlying viability of the plant are

- Raw material: cane juice, molasses, corn — probably primary — credits essential and must be realizable
- Plant size: economies of scale, heat recovery essential
- Investment per unit
- Process used: continuous with recycle, efficient hydrolysis schemes, integrated with other operations

Scale factors are shown in Figure 17. Data are taken from a table in the original paper. The fully loaded costs (ex-plant) as a function of annual output are

	Million gallons		
	10	**50**	**100**
	(units: $/gal)		
Whole grain	1.77	1.33	
Wet milling		1.07	0.97

At the time the article was written, fuel alcohol prices were about $1.65/gal. If a "hurdle

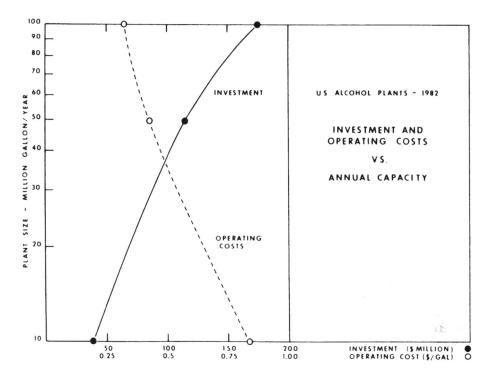

FIGURE 17.   Investment and operating costs vs. annual capacity (U.S. alcohol plants).

rate'' of 40% is required, the selling price would have to be $3.37/gal in the worst case above, to $1.67/gal in the best case. Pro-forma cost statements are given for each of the cases summarized. The analysis leads to the conclusion that very large corn wet-milling alcohol plants might be commercially successful, while smaller plants using less efficient processes have doubtful economic viability. It is recognized that subsidies are required in order to justify bioalcohol production.

One approach to improving economics is given by Oliver.[127] Increasing ethanol tolerance of the producing yeast would improve economic efficiency. A laboratory process was devised to select a culture with greater ethanol tolerance. Rather than submit the continuous culture to a continuously increasing ethanol concentration, a feedback circuit was developed where respiration rate determined ethanol addition rate. Mutants were isolated which could ferment at double the rate of the parent at 10% w/v ethanol content. However, no economic analysis was performed to determine cost impact.

Methods of fermentation optimization utilizing classical response surface statistical techniques (five-level Box-Wilson central composite) were applied to alcohol production.[128] The following parameters were studied and effects quantified: total carbohydrate level, degree of saccharification, glucoamylase dosage, temperature, time. Process response was alcohol yield. The variable with the greatest effect was degree of initial saccharification; time of fermentation has almost as great an effect. Furthermore, there is a significant interaction between these two variables. The relationship between results at varying conditions and economies of investment and operation is discussed qualitatively. The experimental design shows that even with a large number of variables and high degrees of interaction (rather common in fermentation systems), it is possible to generate a response surface which is very useful in predicting performance under different constraints and in determining process and economic optima.

An economic background (especially for Europe) and various options for biological ethanol synthesis are given by Esser and Karsch.[129] In extending the search for ethanol producers to bacteria, the authors find interesting leads that offer economic potential. The search for a thermophilic organism is once again recognized as important to savings in cooling or refrigeration. A thermophile would be rather advantageous in a tropic or subtropic climate. Secondly, search for an organism that would function more or less anaerobically through the entire cycle would simplify mixing and save energy. A number of bacteria were compared to yeast as to substrate range and productivity (published data). Some organisms of interest were *Zymomonas mobilis* and *Thermoanaerobium brockii*. Vacuum fermentation is noted as another potential cost reduction unit operation. Other items which could be studied to improve overall process economics include: continuous fermentation, airlift fermentor, biomass concentration, cell immobilization, genetic manipulation to widen substrate specificity, and improved ethanol tolerance. There are 48 references given.

Actual experimental results using a thermophilic bacteria (*Bacillus stearothermophilus*) to produce ethanol are given by Hartley et al.[130] The authors prepared flow charts and analyzed various schemes to give detailed process economics. One published yeast process was determined to give an ethanol cost of $350/m³. (This corresponds to $1.32/gal.) The experimental ethanol process (thermophile) was determined to give an ethanol cost of about $300/m³ (or $1.14/gal). The economic analyses relate to a "medium-scale" operation producing 120 m³/day of 93.5% w/w ethanol. Costs are in 1980 dollars. Other assumptions are cane at $15/t, steam at $0.35/kg (from bagasse), water at $0.04/m³, chemicals at $0.07/kg biomass and $5/m³ ethanol, stillage treatment at $0.20/m³. Interestingly enough, no by-product credit is taken. Labor, taxes, and insurance are taken at 10% of total fixed capital. Total fixed capital is four times installed equipment cost and capital is costed at 15% per annum.

Total capital costs range from $2.8 to 4.2 million; these figures seem extremely low. Annual operating costs range from $11.5 million for an "ideal, recycle" case up to $15.5 million for the yeast (good practice, modern technology) base case. All test cases involve use of a thermophile operating at 70 to 75°C. The authors note that there may be difficulties with the cost estimation procedure; however, the assessment is primarily comparative so that differences are emphasized, since similar methodologies and factors were used for all test cases.

A relatively novel process design has been presented for significant overall energy savings in ethanol production.[131] The so-called "cold mash system" is projected to save as much as 50% of the energy cost incurred in conventional fermentation-distillation plants. If steam cost is assumed to be $0.071/MJ and electricity cost is $0.0103/MJ, overall energy cost of the proposed, novel process is $0.038/lb. Rather detailed design drawings are included, without capital estimates.

In the analysis of SCP and ethanol ventures via microbial routes, one is struck by the many dislocations that have occurred at a macroeconomic scale. Predictions for usefulness and economic viability have been wide of the mark. A thoughtful political-economic analysis of biotechnology developments has been written by Hamer.[132] The article starts with three well-known predictions that illuminate the point that, "Feasibility studies of biotechnological processing ventures in the bulk and medium volume sectors are usually initiated by ill-defined statements of need..." The three predictions were, first, that Europe and Japan would experience protein shortages for animal feeds from the 1960s onward. Second, fluid fossil fuel prices would rise so dramatically that fuels from alternate, indigenous resources would be produced in many countries. Third, novel products would be needed and produced for use in enhanced oil recovery.

A description of interplay between "global, regional, and national economics" is given, especially as it relates to commodities that are usable as substrates. Then, there is a terse and well-thought-out presentation of how processes are developed, selected, and economically modeled. The conclusion of this section is worth noting, "The domination of economic uncertainty over technological uncertainty is a major factor opposing investment in bio-technological process ventures for the production of both novel and existing bulk and/or medium volume products.'' The statement probably has broader applicability.

There are analyses of ethanol economics, production of biogas, SCP, and biopolymers. Finally, the paper presents some plausible, if far more limited, scenarios where biological processes have a reasonable chance of succeeding. Examples are

- SCP manufacture in countries that have little or no agricultural base, utilizing feedstocks that are difficult or expensive to export
- SCP production from deep earth gas (if found and developed) in countries where climate or other factors mitigate against protein-rich oil seed production
- Biopolymer production in countries that have excess/inexpensive crop production plus partially exhausted oil fields

There is no exhuberance concerning large-scale bioprocessing technologies to supply either food or fuel. National policies which support such ventures may change dramatically and economic viability might vanish. The author concludes by pointing to ''... effective and economic treatment of wastewater and aqueous sludges ...'' as the bulk processing area where biotechnology should and could have a major impact.

Another excellent article (with a good deal of marketing data as well as costing information) by Busche compiles venture analyses prepared in the Life Sciences Division of du Pont.[133] An exhaustive analysis of renewable raw material supply is given. Feedstock prices for corn are compared to gas, crude oil, and ethylene over the last decade. For a 25-million GPY plant (190° ethanol from corn), an investment of $29 million (1980 dollars) is required. Cost plus 30% pretax return results in a $1.42/gal cost (also 1980 dollars). Various biomass costs are derived as a function of raw material type and method of processing. Fermentation conversion cost is plotted against product concentration. Fermentation cost (again including 30% pretax return) can be as low as 5 to 10¢/lb if concentrations near or over 100 g/ℓ can be reached. Finally, product recovery cost is plotted against product concentration. (See Figure 18). Once more the conclusion appears to be that fermentation of renewable raw materials to produce commodity products or feedstocks will become viable at some future time. New developments in both fermentation and recovery processes are essential to restore interest in this technology.

Some laboratory work has been extended to production of organic acids (acetic and propionic) from corn stover.[134] Performance of CSTR was compared to that of immobilized cell reactors (ICR) for production of these acids using *P. acidi-propionici*. From laboratory data, a process design was prepared. A sugar preparation procedure was designed to give a prehydrolyzate containing mainly xylose plus a hydrolyzate containing glucose. Propionic acid is produced from either combined streams or from a single-step hydrolysis procedure; productivity is derived from laboratory data with 1.8 lb propionic acid produced per pound of acetic acid in a CSTR. Separation of the mixed acid aqueous stream is achieved by solvent extraction and distillation. The basis for design is 20 million lb/year (9.1 million kg/year) propionic acid. Design utilizing the ICR was also completed.

The capital and operating costs are detailed in Table 33. Total capital is $11.1 million and the average cost of goods is 36.67¢/kg. If one uses the approximate market price of 73¢/kg for propionic acid and 57¢/kg for acetic acid, the average selling price is 67.25¢/kg. It is likely that a good deal more experimental data, involving larger-scale equipment, will be required before a more complete economic analysis can be performed.

One of the better articles on economic analysis of antibiotic fermentation is by Swartz.[135] With the collaboration of other companies involved in mutation, pilot plant development, and plant design, the author was able to relate yield and rate to relatively simple objective

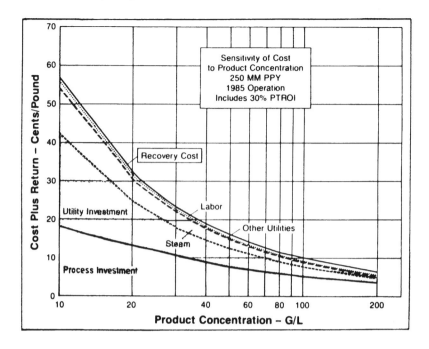

FIGURE 18.   Product (acetic acid) recovery cost sensitivity of cost to product concentration. (Reproduced from Busche, R. M., *Biotechnol. Prog.*, 1, 165, 1985. With permission.)

Table 33
PRODUCTION OF PROPIONIC AND ACETIC ACIDS FROM CORN STOVER (BASIS: 9.1 MILLION KG/YEAR PROPIONIC ACID)

Acid hydrolysis		Organic acid production	
Capital	**$ million**	**Capital**	**$ million**
Feedstock prep	0.55	Reactors	0.67
Hydrolysis	2.05	Acid recovery	0.35
Acid recovery	3.19	Pumps/piping	0.50
Utilities/offsites	2.98	Heat exchanges	0.25
		Contingency (30%)	0.53
	8.77		
			2.30
Operating	**$ million/year**	**Operating**	**$ million/year**
Stover at $20/T	0.70	Sugar feed	3.79
Chemicals	0.58		
Utilities	0.44	Utilities	0.28
Labor	0.59	Labor	0.22
Depreciation	0.88	Depreciation	0.23
Maintenance	0.44	Maintenance	0.29
Taxes/insurance	0.16	Taxes/insurance	0.36
	3.79		5.17

Total acids 14.1 million kg/year
Average cost of goods   36.67¢/kg

Adapted from *Chem. Eng. Prog.*, 80 (12), 63, 1984.

functions for optimization. Then various approaches to process optimization were reviewed and economic impact calculated. A major construction company (A. G. McKee) made available a manufacturing cost breakdown for penicillin G synthesis and isolation; fermentation and isolation flow diagrams are given. One quarter million (5 × 50,000 gal) gallons of fermenter capacity were part of the basis. Since the actual dollar values are some years old, the following percentage breakdown is derived from the data given:

Capital investment penicillin plant (% of total)

Process equipment	24
Installation	5
Insulation	2
Instrumentation	3
Piping	12
Electrical	16
Building	11
Utilities	21
Site	2
Laboratory equipment	3
Spare parts	1
	100%

Engineering	10% of direct installed cost (DIC)
Construction	6% of DIC
Fee	3.5% of DIC
"Excluded costs"	12% of DIC

The "excluded costs" were added by Swartz as an allowance for certain items excluded from the McKee study. It is possible to consider this factor a "contingency", although a contingency is normally included within the DIC (a figure of 10 to 15% is usual). The total fixed investment becomes 131.5% of the DIC. The actual dollar figure was $33.7 million (base year not given, but it was probably 1977 or 1978). In 1985 dollars, that cost would probably be 1.5 to 1.6 times greater. Depreciation (10% on equipment, 3% on buildings), taxes (3% on fixed investment), and insurance (0.7% on fixed investment) are summed to give a total fixed charge. Direct production costs are then calculated based upon the plant design. While exact figures are process, time, and location dependent, normalized figures are useful for comparison to published estimates. Table 34 gives a comparison of various calculated percentages in the Swartz paper to preliminary operating cost estimates given by Black.[136] Final manufacturing costs are calculated for the penicillin process and the normalized figures are given below:

Percent of Total Manufacturing

	Fermentation	Extraction	Totals
Direct cost	51.7	12.5	64.2
Plant overhead	6.3	2.8	9.1
Fixed charges	21.3	5.4	26.7
	79.3	20.7	100.0

Table 34
COMPARISON OF OPERATING COST ESTIMATES

	Calculated from Swartz[135]	Calculated from Black[136]
Raw materials	Determine for each process	Determine for each process
Operating labor	28% of investment	Determine for each process
Direct supervision	37.5% of labor	10—25% of labor
Maintenance	5.3% of fixed investment	2—10% of total capital
Laboratory	4.1% of labor	10—20% of labor
Utilities	Determine for each process	Determine for each process
Supplies	0.7% of fixed investment	0.5—1% of total capital
		Royalties: 1—5% of sales
		Contingencies: 1—5% of direct costs
Depreciation	10% equipment, 3% buildings	5—10% of investment
Real estate tax	3% of fixed investment	1—2% of investment
Insurance	0.7% of fixed investment	0.5—1% of investment
Plant overhead	53.6% of labor, supervision, maintenance	50—70% of labor, supervision, maintenance
		Payroll charges are a separate item at 40% of labor plus supervision

The productivity-cost relationships can then be used to develop very good estimates of economic impact (cost savings) due to successes in process development or plant optimization. The techniques shown are generally applicable. It is important to have these relationships on paper even during early development phases. Even if error is high during preliminary planning and design phases, the equations can be refined in parallel with process development and with focusing of the design estimate. By the time the plant is ready to start up, everyone will have a much clearer picture of where to focus future effort to maximize savings or maximize profitability.

Another article on penicillin fermentation carries penicillin G to 6-aminopenicillanic acid (6-APA).[137] An overall picture of the process is given and capital as well as operating costs are estimated. A logical, stepwise sequence is given which involves a process flow chart, a careful delineation of each step, description of offsites, and listing of all assumptions. A final broth potency (Pen G) of 30.6 g/ℓ is assumed; this is a relatively low value, but does not materially detract from the analysis given. The fixed capital investment is divided into two processing lines, i.e., Pen G and 6-APA. Fermentation and purification (rotary precoat filters and tankage) account for 39.5% of the FCI. Extraction, PenG recovery, and 6-APA production account for 11.1% and solvent recovery, offsites, and other buildings account for the remaining 49.4%. Although possibly a bit high, it should not be surprising that almost half of FCI relates to "auxiliary" process functions. As the authors note, this points clearly to the need for a multiproduct plant where overhead and fixed costs can be shared. With a 20% contingency, the total FCI (March 1983 costs) is $25.1 million. Output is 275 tons/year of 6-APA, which is estimated as 5% of the overall world market. Operating costs were calculated as:

Variable	$30.27/kg 6-APA
Fixed	22.30
Total	$52.57/kg 6-APA

The variable cost segment includes $22.12/kg for raw materials and $8.15/kg for variable utilities. The fixed cost includes $6.10/kg for depreciation and $6.28/kg for operating labor (79 operators, including maintenance). An interesting part of the article concerns cost reduction methods with some quantitative analysis of impact. Some are obvious, but the approach is applicable to any fermentation facility. Cost reductions can be affected by:

1. Improvement in fermentation rate and yield
2. Improvement in recovery yield
3. Lower raw material costs (or in-house production)
4. Substitution of less costly raw materials
5. Increased production with constant labor supply
6. Use same labor force to produce additional products
7. Work force reduction (improvement in process or controls)
8. Multiple product synthesis with little or no additional investment
9. Combining processing steps

For the facility described, certain potential cost reduction studies were not mentioned. Investigation of alternate use of waste streams (animal feed, fertilizer) or alternate waste stream processing might offer appreciable savings. Energy cost reduction (cogeneration, heat recovery) provides another area of potential savings.

An excellent article on commercialization potentials for plant tissue culture includes a number of economic considerations.[138] Not only are there colors, fragrances, and flavors that have commercial interest, but many actual and potential herbicides, pesticides, and pharmaceutical products may be derived from plants and plant tissue culture. Plant cells grow rather slowly compared to bacteria or fungi. Plant cells may have doubling times of 20 hr, but very often the doubling time may be as long as 100 or 200 hr. A bacterial fermentation cycle may last from less than a day to a few days; a mold or fungal fermentation may take a few days to a week. A similar cycle for plants can be calculated at 0.5 to 3 months. The authors have made these assumptions and presented theoretical cost estimates (continuous immobilized cell process):

Capacity	20,000 kg/year	Production costs include: raw
Doubling time	2.5 days	materials, labor, supervision,
Cycle time	15 days	utilities, depreciation (10
Cell density	20 g/ℓ	years), maintenance, taxes, insurance; indirects at 15% of capital

Product conc, %	0.1	1.0	10
Fixed capital, $ million	340	84	21
Production cost, $/kg	5,900	1,045	228

Sucrose accounts for almost 80% of the total raw material cost; molasses use would significantly lower costs. The analysis indicates that, at production levels currently obtainable in plant culture, high value products with medium or low volume markets should be targeted for commercialization. They also present various scenarios for rose oil production via batch fermentation. Based on annual production (730 to 3650 kg/year), production costs vary from $542 to 1215/kg. Fixed capital investment is low ($3 to 7.9 million) and payout time varies from 1.4 (high output) to 2.9 years (lowest output). Corresponding ROI is 53.5 to 22.8%.

Utilization of immobilized whole cells offers numerous advantages and prototype systems are described. The immobilization technique offers increased cell density, higher reaction rates, uncoupling of growth and product formation, improvement in specific rate, reuse of biocatalyst, and continuous processing. It is likely that at operating capacities of 1×10^5 to 1×10^6 kg/year, production costs of plant-derived substances could fall in the range of $20 to $25/kg.

Fuller[139] details some of the history of plant cell culture and does not shrink from detailing some of the flawed logic that went into extensive effort to produce specialty items from plant cells. Technology was applied even though conditions were far from optimal. Surprisingly few processes can be considered "successful" at even pilot scale; the author considers shikonin production the only one that can be called commercially feasible. Many references are given to recent work involving phytoalexins, polyacetylenes, alkaloids, and quinones, among others. It is too early, apparently, to make a meaningful assessment of economics. Minimal prices per kilogram of product have been calculated to range between about $20 to $2000. If, however, the product needs little downstream processing and can be sold as a cellular entity, the breakeven price is reduced, perhaps, by one or two orders of magnitude. It seems that the volume of material required per year dictates a multifunctional plant where sequential scheduling can produce numerous products; another economically attractive alternative is to produce a single cell mass and use this as starting material for separation and purification of many related chemicals. The author concludes that the large resources invested in engineering technology may be misplaced; time required to find and develop a process and ultimate profitability cannot be guaranteed with any level of certainty.

Integrated Genetics has worked out a scenario for albumin (a plasma constituent) via recombinant DNA synthesis.[140] Assume a bacterial system gives 10 g cells per liter. Further assume the cell is 20% protein and 40% of this protein is desired albumin. Therefore, product yield is 800 mg/ℓ. If the cost of production (to the broth stage) is $1/$\ell$, crude product cost for a 12.5-g dose would cost $15.60. Concentration, purification, and packaging of the albumin would raise the cost well beyond the (early 1985) selling price of $30 to $35 per 12.5-g vial. The company will not pursue the venture.

Chapter 4

PLANT CONSTRUCTION AND REGULATORY AFFAIRS

I. REGULATORY CONSIDERATIONS

Both the time and cost expended in gaining regulatory approval for a new bioactive compound are very high. One source estimates that drug-maker development costs are running from $80 to $100 million per new chemical entity.[141] Some efforts in 1984 to 1985 may cut the "normal" 30-month interval for FDA approval of a new drug application (with documentation that may run to many tens of thousands of pages) by 6 months.

The situation is not limited to the drug industry. Biotechnology, in its broader sense, has many points of contact with the food processing industries. Areas of biotechnology that impinge upon food processors include: genetic manipulation of plant cells, pure culture fermentation, single cell protein synthesis, aquaculture, plant/animal cell culture, and enzyme biosynthesis and use. Clausi[142] notes that food companies sense that competitive survival may depend upon an understanding and utilization of these emerging technologies. Yet costs are high, capabilities are uncertain, economic impact is ill defined, and last but not least, the regulatory picture is cloudy. It will be necessary to establish a business framework for transformation of scientific findings into economically viable industrial applications. One must understand that the nature of discoveries and their potential broad interactions in nature point toward the need for a new regulatory framework, as well.

Korwek notes that "obtaining regulatory clearances to manufacture and market products often involves millions of dollars and enormous expenditures of resources. Regulatory hurdles can result in long delays in delivery of useful products and drive up consumer costs."[143] He discusses all the legal framework for regulation of biotechnology and specifically reviews FDA authority:

- Relative to foods and food additives (including history and definition of GRAS)
- Over human and animal drugs (premarketing clearance and approval for changes in manufacture)
- In regulation of new methods of manufacture of medical devices (definition of Class I, II, III, and specific requirements)

Korwek believes that a scheme already exists for regulation of food ingredients, drugs, and devices and any technical problems or regulatory issues raised by "biotechnology-produced" are not peculiar to the technology. A key point is made in that industrial applications of biotechnology involve traditional methodologies. Therefore, history and prior practice are relevant in evaluation of the safety of products resulting from novel technologies. The main burden of defining safe and suitable, or "identical", or product purity resides with the manufacturer. There is a final plea not to introduce unneeded restrictive or excessive regulation which will hamper commercial development.

However, a general consensus on risks of biotechnology applications does not exist. While mutation/selection methodology on a more or less rigorous basis has been employed in industrial fermentation for some 40 years, the "quality" and "quantity" of the contemporary genetic changes appear to elicit a different response.

One consensus[144] is that technicians involved in genetic manipulation of micoorganisms and crop plants are not well versed in ecological principles and are not considering potential environmental risks. The view that genetically modified species will be readily eliminated in nature (or outside a carefully controlled environment) has not been universally accepted.

The potential for gene transfer between herbicide-resistant crop plants and weeds was noted; eliminating the altered weed would become a horrendous task. In the same article, note is taken of a "...stinging attack on the biotechnology industry". A congressional staff member delivered the attack while, unsurprisingly, calling for new and more severe regulations on the industry.

The FDA recognizes that new technologies require more than minor readjustments to its mission and organization. Biotechnology is recognized as responsible for an "...ongoing revolution in medical and food technology".[145] While major emphasis will be placed upon drugs, vaccines, and gene therapy, it is recognized that equal emphasis will be placed on modified plants for human consumption. While not a new concept, postmarketing surveillance will be essential and will probably be expanded. Programs are not fully developed in this latter area of ongoing research. One can estimate that these programs will be complex, lengthy, and costly. Regulation will become ever more active and more extensive.

It is not merely the FDA that is involved; regulation of the biotechnology industry will reside within other existing organizations. The FDA, the Environmental Protection Agency (EPA), and the U.S. Department of Agriculture (USDA) are considered sufficiently large to cover all regulatory concerns. While jurisdictional disputes (not unknown today) are expected to continue, it does not appear that a new or super agency will be required. This may not be true for current statutes; one can expect major changes with time. The federal government is taking the view that genetically engineered products (whether in health care, chemicals, or agriculture) do present different and possibly novel risks. Therefore, regulation will occur on a "case-by-case basis". While such an approach may present an occasional pleasant surprise, it is more likely that greater uncertainty and higher cost will be the usual result.

The policy proposal prepared in late 1984 by the Office of Science and Technology Policy (OSTP) attempts to codify pathways to be used in regulation of bioengineered products.[146] This document is useful for many reasons. There is a complete listing of "biotechnology authorities". Each individual guideline or authority is detailed with affected process or product shown, agencies involved, and various cross references. Separate headings are given for:

1. Licensing and other premarketing requirements
2. Postmarketing requirements
3. Export controls
4. Research and information gathering
5. Patents
6. Air and water emissions
7. Requirements for federal agencies

Note that all products of biotechnology are covered. A useful glossary of terms is added in order to "...assist the reader". The usual proviso is given so that the terms are not legally binding on any federal or nonfederal agency or organization. Projection from the general to the particular is probably hazardous.

The proposed policy, insofar as it relates to food products, is reviewed and analyzed by Middlekauff and McCue.[147] The article does include a useful, short history of U.S. food laws. The authors conclude that, "by deferring the basic regulatory issues of genetically engineered products until they are presented to the agency in individual products, FDA has permitted itself more freedom than would otherwise be the case." The greater freedom can easily be equated with greater uncertainty for the petitioner.

In 1985, OSTP established the Biotechnology Science Coordinating Committee (BSCC). The purposes of this committee are to:

- Serve as a coordinating forum and develop a consensus on evolving problems and solutions
- Promote consistency in federal agencies' review procedures
- Facilitate cooperation among federal agencies on emerging scientific issues and technologies
- Identify gaps in current scientific knowledge

It is unclear whether this committee will aid in clarification of issues. The objectives are noteworthy and do relate to gaps that exist in the existing regulatory framework.

A USDA official has written a summary of the interactions and guidelines insofar as *food* products resulting from biotechnology are concerned.[148] All relevant agencies and NIH guidelines for federally funded rDNA work are reviewed. This article is very useful since it not only discusses relevant statutes, but gives detailed references to relevant portions of published guidelines, orders, laws, and regulations. Further, pending and completed litigation is discussed.

The same author has prepared a more lengthy proposal for a "science-based regulatory model" which can be applied to products of gene transfer in food organisms.[149] He notes and comments on the existing pertinent legislation. Included are

> Federal Food, Drug, and Cosmetic (FD & C) Act
> Federal Meat Inspection (FMIA) Act
> Poultry Products Inspection (PPIA) Act
> Egg Products Inspection (EPIA) Act
> Virus, Serum, and Toxin (VSTA) Act
> Toxic Substances Control Act (TSCA)
> Federal Insecticide, Fungicide, and Rodenticide Act (FIFRA)
> National Environmental Policy Act (NEPA)

There follows definitions and explanations of components of genetic manipulation in microorganisms, plants, and animals. The regulatory model proposed is based upon research and development as divided into three phases; that is, delivery, expression, and commercial. During the first phase, genetic material transferred from a source is exogenous to the host organism. The need for careful monitoring in this phase is noted. Environmental impact of release of the vector or vector/gene combination may be adverse; in any event, it is an arguable issue. The second, or expression, phase begins with incorporation of new genetic material into the genome of the host. During this phase, the host organism is considered a test or experimental organism. Controlled expression must be shown to occur in the test organism. In the final or commercial phase, the transgenic organism is commercially developed for food use. Each of the phases of the model are separated by discrete and technologically achievable events. Further, the phases are roughly correlated with existing statutory authorities of federal agencies most involved in development of food biotechnology. Here is another attempt to utilize an existing framework and not present new and untried barriers to successful implementation.

Another aspect of regulatory concern, which has been noted earlier, involves release of genetically manipulated microorganisms. More specifically, the concern relates to products of recombinant DNA technology. Korwek and de la Cruz discuss environmental releases of this type and applicable legislation.[150] The authors note that if the broadest definition of the new biotechnology is used, almost an infinite number of potential environmental interactions are possible. The examples cited are

1. Creation of new plant varieties that contain a more desirable amino acid pattern or that are resistant to herbicides, pesticides, and climatic extremes

2. Creation of modified domestic food animals for improved or modified milk or meat production
3. Genetically altered microbial insecticides
4. Microorganisms designed to treat liquid or solid wastes or to degrade highly toxic compounds
5. Organisms to leach metals from mines and/or concentrate metals from dilute solutions
6. Use of microorganisms in direct injection into substrates or wells for enhanced oil recovery

Note that these examples concern modified organisms designed for release into the environment. That is, containment is not part of the functional design. Production of bioactive entities via rDNA technology could conceivably involve containment (both fermentation and extraction/purification), but costs would be higher by an order of magnitude. Most fermentation products made today do not involve total containment of effluent air, waste streams, or extraction apparatus. Some discharge of the producing microorganisms into the environment occurs in almost every case. It is probable that in rDNA processing to date, similar discharges have occurred. As the authors note, however, predicting the type, scope, or severity of environmental effects associated with deliberate (or even accidental) release of genetically altered organisms may be difficult, if not impossible. The regulatory authorities and history are detailed; requirements for an environmental impact statement (EIS) are reviewed. Acts or agencies that may have oversight are

1. National Environmental Policy Act (NEPA)
   A. Council on Environmental Quality (CEQ)
2. The Clean Air Act (CAA)
   A. National Ambient Air Quality Standards (NAAQS)
   B. New Source Performance Standards (NSPS)
   C. National Emission Standards for Hazardous Air Pollutants (NESHAPs)
   D. State Implementation Plans (SIPs)
3. Federal Noxious Weed Act (FNWA) under USDA
4. Plant Quarantine Act (PQA)
5. Federal Plant Pest Act (FPPA)
6. Resource Conservation and Recovery Act (RCRA)
7. Comprehensive Environmental Response, Compensation, and Liability Act (CERCLA, also known as Superfund)
8. Clean Water Act (CWA)

After reviewing the coverage, including overlaps and gaps, the authors conclude that:

> Extant authority should be utilized to develop a rational federal regulatory approach in order both to preclude the enactment of conflicting or inconsistent state and local laws, and to maintain the competitive edge of the United States in the biotechnology area. Unreasonable restrictions or prohibitions on deliberate releases provide no absolute guarantee of safety, especially in light of the continuing development of biotechnology industries in other countries.

Given recent history in this area, it is not likely that the orderly and rational approach will be followed. More likely are more court cases and more legislation.

In early 1986, a report was issued by the Biotechnology Study Group organized by the EPAs Science Advisory Board.[151] The subjects were direction of EPAs research program in biotechnology and identification of gaps for performing risk assessments on products of biotechnology. The suggested program, if enacted, would make EPA the lead governmental body for research on environmental effects of bioengineered products and their production.

A new law is suggested which would establish a new program that would review proposed releases of genetically modified organisms. Each situation would be reviewed on a case-by-case basis. The program would also coordinate research efforts to better understand organism survival, genetic transfer *in situ,* dispersal, health effects, environmental effects, and possible remedial action. *All* modified organisms would be covered — plants, animals, microbes. Not unexpectedly, major grants are suggested for academic (and other) researchers to prepare a reliable ecological assessment model that could predict an organism's fate after release.

These issues are probably just as acute for the food industry. Whereas the first recombinant DNA products are drugs (as insulin, interferon, growth hormone, vaccines) and major technological barriers exist in plant and animal modification, the regulatory issues are still there. From a certain perspective, one could conclude that the barriers may be greater in the area of food technology. Whether or not existing laws are sufficiently flexible to encompass coming changes in agriculture is arguable. Gibbs and Kahan[152] discuss the real and potential hurdles in regulation of biotechnology as applied to food. One key point is that any GRAS (generally recognized as safe) substance that is genetically altered, by any means, may have its continuing validity under GRAS reviewed by the FDA. The FDA can determine whether the material (plant or animal) has been "significantly" altered. That is to say, there is no statutory definition of a "significant alteration". The labeling question falls in the same category. It is unclear when (or if) genetic alteration must be described.

"Sauce made from genetically modified tomatoes" may prove an advertising boon; on the other hand, it may have a completely negative connotation. The entire issue of "standards of identity" falls in this area. Defining whether or not a plant or animal (or even a product of enzyme activity) fits the extant standard of identity after genetic manipulation of the living organism (or the cell which produced the enzyme) presents a host of problems. It is likely that environmental impact statements will be requested prior to introduction of a recombinant plant or animal. Given environmental releases or their publicity in 1986 (see below), requests for such statements will become routine. Rather than face court challenge, regulatory agencies may simply acquiesce and make the request of the petitioner.

It is understood that clearance for any existing food additive (with very few exceptions) does not identify any specific method of manufacture. As long as good manufacturing practice is followed, it is the product and its characteristics, rather than the process and its characteristics, that are regulated. It would seem that use of recombinant technology should not trigger a new filing of a food additive petition. It is well known, for example, that many new *mutant* (not programmed DNA alteration) cultures have been introduced for preparing vitamins or amino acids by fermentation. No new food additive petitions have been filed. However, the FDA has indicated that recombinant DNA technology will trigger a new filing. It may very well be that microorganisms are already in use (for the fermentations mentioned, or others) that have been derived from recombinant DNA techniques with no subsequent new filing.

It is safe to say that appropriate regulatory bodies will maintain their vigilence over the food supply. The term "recombinant DNA" or "genetic engineering" as applied to food substances will create complications in the regulatory review. More time and more money will be required to overcome regulatory interest and concern.

In early 1986, the General Accounting Office sent a voluminous report to the House Committee on Science and Technology which stated that the USDA policies on regulation of biotechnology need more clarification. There were suggestions made as to how the regulatory structure could be improved and as how a new effort to communicate with the public could be made. Reference was made to intra- and interagency squabbling and unclear lines of authority. The GAO report did make the required disclaimer that there was no desire to impose or create cumbersome regulations that would stifle growth and development of biotechnology; however, suggested studies, reorganization, reassignment, and new procedures would effectively hamper clearance. This would impact short- and medium-range projects and products in the agricultural area; there might be a long-term benefit, but that would be conjecture at this point in time. The GAO report is another indication that the nature and potential application of the new technology is different both qualitatively and quantitatively from past changes in the scientific arena. Existing regulatory bodies may have the inclination, infrastructure, and knowledge to handle the new biotechnology and its products. To date, there has been little manifestation that successful integration into existing structures can occur with little or no dislocation.

The first known regulatory confrontation which involved an environmental release of a recombinant organism concerns a test performed by scientists at Advanced Genetic Sciences. In this instance, modified bacterial strains were injected into trees outside a laboratory or controlled greenhouse environment. The company was fined and the EPA suspended their permit to field test, but a great deal of negative publicity was generated. In a short period of time, a second regulatory issue surfaced. In early 1986, the USDA approved an animal vaccine produced by Biologics. The vaccine is produced from a genetically altered virus. Mr. Jeremy Rifkin, of the Foundation on Economic Trends, who has been following the evolving technology for years, charged that in-place guidelines were not followed. In any event, later in 1986, the USDA suspended the license. After a review, the license was reinstated. These two events do not bode well for future regulatory clearance. It is probable that there will be a further development of a "bunker" mentality which will incline towards requests for more testing, more paperwork, and more delay. It is not difficult to predict that regulatory approval will be delayed and commercialization will be slowed.

Actual data on cost of drug approvals is not easy to come by. Estimates range from tens of millions upward to $100 million (as noted earlier). Certainly, many years are involved in gathering and organizing relevant data. In some instances, more than 10 years have been required. Daly[153] notes in a recent book that new drug clearance may take up to 7 years and cost as much as $70 million. This is true even if the rDNA product is identical to an existing one.

As for a food additive petition, one attorney involved in this endeavor estimates cost of approval of a food additive petition ranging from $100,000 to $1 million, depending upon whether chronic feeding studies are required.[154] The final determinant as to extent of analysis and extent of chronic feeding studies rests with the FDA.

Some information on medical devices was also given in the same communication.[154] "Approval" of class I and II devices via the 510 (k) route is believed to cost under $2000. Documentation required normally runs under 50 pages.

A premarket approval application (PMA) for a class III device is more time consuming and costly. An estimate of such cost is several hundred thousand dollars.

While market potential, as well as commercial advances, tends to magnify the importance of U.S. regulatory bodies and their deliberations, a world-wide regulatory approach has been under review for many years. The Paris-based Organization for Economic Cooperation and Development (OECD) created an ad hoc Group on Safety and Regulation in Biotechnology in late 1983.[155] The U.S. is a member (also Japan and many western European countries belong). The study group centered upon agricultural and environmental applications of recombinant DNA technology. Principles were agreed to for applications of these organisms and it is hoped that some internationally agreed-upon guidelines can be established. It is abundantly clear that national boundaries may be insufficient to contain a microorganism or plant that has been released in one state. It is also clear that national sovereignty will not be quickly surrendered in establishing regulatory guidelines. While such an international group does perform a useful function, and the resulting report may serve as a useful guide, it is difficult to see how mutual agreement on a governmental level can be achieved rapidly.

A number of actual, as well as potential, problems in food safety are discussed by

Knudsen.[156] He stresses the importance of international deliberations (specifically noting the OECD study), since genetically modified organisms "... will eventually be dispersed worldwide and would not recognize human-made trade barriers or borders". A listing of food safety aspects to be considered relative to gene technology is given.

One can safely conclude that the regulatory picture in the U.S. for novel bioengineered products prepared by rDNA techniques is somewhat confused and is in a state of flux. Federal regulatory agencies have not yet occupied appropriate spheres of responsibility. The legislative picture is incomplete, but draft legislation has been prepared in a number of congressional committees. It seems that the overall approach will be pragmatic or empirical. Clearance and permission will be granted on a case-by-case basis until a sufficient number of precedents are established. Blanket coverage will not be given for years, if ever. Incremental progress can be expected in the regulatory process; advances in science will probably outstrip the regulatory scheme. This entire scenario, slow and complex as it is, is based on a major premise; that is, there is and will be no serious environmental upset that can be blamed on rDNA technology. This proviso has already been breached even though no serious consequences were noted. Should such an upset occur, one can predict a profusion of laws and regulations to further inhibit technology transfer. The burden of "proving" the technology and allowing on-going progress rests with those organizations who have received regulatory clearance (and those who are now applying). Sufficient controlled testing must be completed by the initiator to insure safe environmental release.

A somewhat more constrained area of regulatory concern involves good manufacturing practice (GMP). Both drug and food manufacture are closely regulated by the U.S. government. Pharmaceutical manufacture is covered by a set of guidelines called "current good manufacturing practices" (CGMP) which has as its aim the guarantee that finished pharmaceuticals conform to preset standards of identity, potency, purity, and quality. An analysis of CGMP is given by Harrison.[157] He notes that firms planning to manufacture pharmaceutical products should distribute relevant sections of the Code of Federal Regulations (CFR) to as wide a group as possible, as early in the planning phase as possible. This can be extended to any production process (including distribution and marketing) that involves biologics, drugs, and foods.

A review of the important sections and their headings in the CFR will indicate the importance of the document.

Title 21 CFR

Part
110 Umbrella GMP
 21 Drug GMP
820 Medical device GMP
211 GMP — finished pharmaceuticals
225 GMP — medicated feeds
226 GMP — medicated premixes
606 Blood and blood components

A further subdivision of CFR 21 Part 110 (revised as of April 1, 1985 with the full title, "Current Good Manufacturing Practice in Manufacturing, Processing, Packaging, or Holding Human Food"):

Subpart

A   General provisions
B   Buildings and facilities

C	Equipment
D	(Reserved)
E	Production and process controls

The FDA published revisions to the umbrella GMP as a final rule issued June 19, 1986 (CFR, Vol. 51, No. 118, 22458). The document covers Parts 20 and 110. Part 110 retains the same subparts, with subpart F reserved and subpart G entitled, "Defect Action Levels". The final rule is effective as of December 16, 1986. There are no changes which can be related to developments in biotechnology.

Part 211 has even more subdivisions:

Subpart

A	General provisions
B	Organization and personnel
C	Buildings and facilities
D	Equipment
E	Components and drug product containers and closures
F	Production and process controls
G	Packaging and labeling control
H	Holding and distribution
I	Laboratory controls
J	Records and reports
K	Returned and salvaged drug products

It is important to note that the firm that holds the approved NDA (new drug application) is responsible for final quality control even if the drug is manufactured under contract. Simply put, final quality control cannot be delegated. Whereas new drug approval is a tedious and long-term process, there is a reasonable end in sight. Adherence to GMP is an ongoing process. Inspection and reporting is ongoing and knowledge of GMP must be maintained as long as the product is being manufactured or being distributed. As might be expected, there are also regulatory requirements for clinical, nonclinical, and toxicology laboratories involved in testing and characterization of drugs. These are called GLP or "good laboratory practices". Strict requirements exist for data collection and all accompanying practices.

II. PATENTS

As part of the regulatory picture, it is important to consider the patentability of a microorganism or process. One cannot limit the analysis to one country; the biotechnology scene is international in scope and single country coverage is not adequate in most circumstances. The patent issue as it relates to biotechnology is not only complex, but involves voluminous commentary. There will be no attempt to give a legal background or legal status for biotechnology inventions; however, certain major items will be noted and references given for a more thorough study. A cursory review of biotechnology patent litigation (some of which is ongoing) for the past decade will show that economic impact and potential for gain or loss are both vast.

In 1983, there were 734 biotechnology patents issued by the U.S. Patent Office, according to one authority.[158] In 1984, there were 826. While patent laws apply to all technologies, biotechnology patent law may involve deposit requirements. This occurs because all patents have these legal requirements:

- A written disclosure is essential.
- Description outlines manner and process of performing or using the invention.
- Sufficient clarity is used so that any person skilled in the art or science would be able to practice the invention.
- "Best mode" of practice by the inventor must be disclosed.

If a specially modified microorganism is integral to the practice of the invention, it must be deposited at an accessible location. Where products of rDNA effort are involved, depository requirements are clear. Patent procedures of many countries require a sample of the microorganism to be deposited with "a recognized third party" that will prepare and maintain a supply of the culture at least for the life of the patent.[158] There are at least a dozen such authorities around the world.

An interesting international survey of patent practices and comparative protection in the biotechnology field was completed in 1985.[159] The authors note that there were more than 600 industrial firms in the U.S. alone (end of 1983) which used or produced biological molecules or microorganisms. Fermented food and beverage industries were excluded. These companies (and others in the chemical and pharmaceutical industries) have been the major supporters and users of the patent system. This will not change in the near future. The monograph attempts a comparative review of existing patent laws among many countries; emphasis is on biotechnology applications. The authors also interpose independent opinions on the usefulness and adequacy of existing law and comment on possible needed changes.

There is a useful introduction which covers exclusions from patentability, the objectives of the patent system, basic principles of patent law, and development of such law (including differences) in various countries. It is noted that in the U.S., cost of developing a new drug would range from $50 to $100 million and require many years. An imperfect patent system, or one that does not foster international exchange, would act as a major detriment to development. A questionnaire approach was used (34 questions) and responses were grouped and analyzed. Responses came from responsible government offices in contacted nations. Individual country practices and procedures are described. As one might expect, procedures and requirements vary. It would be a relatively simple matter to file and/or deposit in one locale and lose the opportunity to file elsewhere. Timing and exclusionary rules vary from country to country and even after the decision to file is made, great care must be taken to follow sequences (especially as far as deposition is concerned) that will allow complete protection around the world. Requirements for deposition of a microbial, plant, or animal cell strain mean that needs go well beyond those of a conventional patent. The authors summarize responses on this matter and point to a "... practical solution as entailing the following":

1. Release only where enforceable rights exist
2. If above not possible, restricted release to an independent expert
3. Destruction/return of strain if no patent issued
4. Restrictions on release to patent-free countries via
   - Nontransmission to any other person
   - Prohibition of export
   - Release only to residents of countries where patent rights exist
5. Released culture to be used for experimental purposes only
6. Release restrictions to be extended to derived cultures
7. Burden of proof on third party who obtained culture should breach be noted

Patent office practice in each country is discussed as are attitudes toward the patent system. Specific recommendations are given for improvements and the need for international harmonization is stressed. It is likely that the current dislocations and differences will persist and the discoverer would be wise to exercise extreme caution in pursuit of world-wide patent protection.

Patent pitfalls (or a few of them) are discussed by Benson[160] in a short article. He covers recent cases that involve willful infringement (and associated multiple damages), patent term extension, proposal legislation on international trade in biotechnology products and patents, the need for adequate search and evaluation (prior publication), deposit requirements, and other unexpected legal phenomena. The complexity of the science is easily matched by the rapidly changing legal framework. The value of microbes or tissue cultures as well as the value of the products, coupled with a laboratory race to "express" or "produce", leads to a situation where litigation is stimulated. An error or misunderstanding (even under the best of ideals involving scientific cross-fertilization) can lead to very costly legal proceedings and even more costly legal judgments. The daily newspapers may not present the scientific or legal issues in a sufficiently detailed fashion, but the dollar signs and the diseases involved stimulate a great deal of public interest.

The same subject, on an international perspective, is discussed by Lawrence.[161] It is clear that there is more than one option available for commercialization of an invention. Unprotected publication or dissemination is one option. Secrecy is one corollary; one would depend on maintenance of control through confidentiality without patent protection. Another involves secrecy with licensing of knowhow. Patent protection is yet another option. The author notes that not every invention (even if exciting or of economic impact) must or should be patented. Each case must be reviewed and decided on its own merits. He presents his own guide on the issue of "new and not obvious". If there is an established commercial problem that people have tried to solve, then an effective solution to the problem is probably not obvious. This is also true for a hazard (disease state, malformation, improper development, and so on) that people have tried to avoid, cure, or resolve. Each step in discovery, description, sample release, and definition of "desired monopoly" is discussed. The last step is extremely important, since it defines the essential feature of the invention and the protected field granted to the applicant. A framework of patents, even with some blank areas or weak links, can often offer a great deal of leverage in creating a monopoly or in licensing technology. The strategy of patents is a critical determinant of business success. Above all, one must be willing to protect the "monopoly" and if the occasion arises and infringement suspected, suit must be brought. Defense of the property of knowledge is often necessary.

The entire subject of depository requirements is discussed by Hofer.[162] After describing what a patent is (and is not), the author discusses the famous *Diamond v. Chakrabarty* case of 1980. In this case, a 5-4 decision of the U.S. Supreme Court allowed for claims to compositions of living matter as being patentable subject matter. It is noted that as late as 1984 there was a small group in the Patent and Trademark Office that handled some 600 cases in this area. By 1985, there were two expanded groups involved with over 3000 cases.

The issue of enablement as related to genetically engineering biological entities is extraordinarily complex. Since the science is evolving, what is the capability of one "skilled in the art"? As part of the enablement requirement, actual availability of one or more critical starting materials was considered a key part of satisfying the legal requirement. The deposit of the living tissue, or part of the living tissue (as a plasmid or even part of a plasmid), on or before the filing date would satisfy a great portion of the enablement requirement. Of course, one solution has a tendency to lead to host of new problems or, as some might say, new opportunities. Deposit requirements would be difficult to satisfy under any circumstances, due to the nature of the material to be stored. Adding timing and secrecy requirements along with rather complex directions for maintenance, regeneration, or use testing creates a very involved situation that is fraught with new and difficult hazards. There are many expedients in use and many unresolved issues that exist in this area in the U.S. There may be other, equally complex issues for foreign depositories. As Hofer notes, "The creation of new products using deposited materials as critical raw materials raises a number of infringement/experimental use issues in addition to grave questions regarding ownership if not inventorship." Costs incurred by one U.S. depository, The American Type Culture Collection (ATCC), are given:

- $145 per U.S. deposit for first year
- $100 per deposit per year thereafter until issuance
- $570/deposit for extension for foreign filing, non-Budapest with prior U.S. fees possibly credited
- add $100/deposit to meet Budapest requirements
- $300/deposit for reporting of sample requests

The entire issue of when to patent, when to deposit, and how to achieve return of deposit (abandon patent or nonissue of patent) is discussed. Given the nature of legal barrier complexity or lack of same from one country to another, it is easy to envisage a situation where deposition leads to improper transfer or even theft of a very important item of deposited living tissue. The complexity of foreign filing and its pitfalls are also reviewed. The patent system is not a failsafe system, as court calendars show, and the added burden of depository requirements helps to confound the issue. Each case must be reviewed carefully both for U.S. filing and for world-wide coverage. Potential gains and losses must be weighed; there is no simple or general rule to follow.

The Drug Price Competition and Patent Term Restoration Act (1984) was passed to allow extensions of up to 5 years for patents related to drugs, food additives, and medical devices approved by the FDA. The patent extension provisions of the law are to be jointly administered by the FDA and the U.S. Patent and Trademark Office. Products that may fall under the patent extension provision must represent the first commercial use of the chemical or device. A very important proviso is, however, that even if FDA has previously approved a product, a duplicate entity made by rDNA technology can still be eligible for patent term extension. This is a unique concession to a novel technology. The other side of the coin is that the patent extension can be applied for because of regulatory delay in clearing the "duplicate" chemical. The owner of the patent must apply for a patent term extension within a 60-day period beginning on the day of product approval. If this relatively short interval is missed, the extension will not be granted.

While the rDNA exception is not involved in the following cases, six drugs have had their patent term extended (early 1986) under the Patent Term Restoration Act in order to make up for exclusive marketing time lost while FDA studied the new drug application.[163] The drugs and companies involved were

Ridaura	Smith-Kline Beckman
Sectral	Wyeth
Seldane	Merrell Dow
Tornalate	Sterling Drug
Tonocard	Merck
Promit	Pharmacia (Sweden)

The impact of this 1984 act (abbreviated as DPC-PTR Act) is discussed by Ryan.[164] He believes that the act "... creates opportunities for developing new drug therapies and for expanding the market of existing products by providing incentives to research-based companies for innovative research and to both research-based and generic-based companies for new delivery systems and market extension." It is important that all company employees

Table 35
TYPICAL FOREIGN PATENT FILING FEES

	U.S. dollars		U.S. dollars
Argentina	1750	Italy	1200
Australia	550	Japan	1300
Belgium	1000	S. Korea	950
Brazil	900	Mexico	1050
Canada	400	New Zealand	400
Columbia	1450	Norway	1800
Denmark	1850	Poland	1150
Egypt	900	Russia	950
France	1550	S. Africa	400
Germany (E)	1350	Spain	900
Germany (W)	1200	Sweden	2150
Great Britain	400	Switzerland	1850
Holland	1350	Taiwan	700
India	125	Turkey	1000
Ireland	500	Venezuela	1500
Israel	400	Yugoslavia	850

are aware of the act and its significance; otherwise strategic opportunities might be lost. Provisions of the act are complex, burdensome, and invite competition, but opportunities also are presented. New issues in licensing have emerged and strategies relating to infringement have altered. The major burdens of the act will probably fall on corporate legal and regulatory staffs; however, the selection of projects and marketing programs will also be heavily impacted. Therefore, an ongoing task force approach is needed to maximize potential advantages offered by this legislation.

A useful review article on patents in the fermentation area (at least up to the late 1970s) was written by Marcus.[165] Emphasis is on fermentation processes and products. The examination (Patent and Trademark Office) is covered in great detail. Patent practice and problems are reviewed; deposit requirements are covered. It is abundantly clear that as science and genetics change, the legal and regulatory framework will be modified to reflect the evolution (both figuratively and literally). Patent office response to changes in technology often lag the scientific discovery. The potential for loss of novelty and accompanying financial loss is very high. The speed of international communications, coupled with the complexity of various national laws, means that attorneys (in-house or in an outside firm) become essential to successful financial development of a novel microbial process or product. The choice of when and where to seek patent protection (or when or where *not* to) must be made after a thorough technical-legal review of pros and cons. This has always been true in most technologies; it is even ''more'' true in a rapidly evolving technology. There are many scientists with graduate degrees in microbiology or biochemistry who have opted for a law degree as well. These persons are now thoroughly immersed in the regulatory-patent arena involving genetic engineering. It is a safe prediction to state that the numbers of such multidegreed persons will increase in the years to come along with products and processes that result from application of this technology.

Filing a U.S. patent application is costly even after the patent itself has been prepared. The added costs include basic fees for the local associate and government filing fees. The associate will normally prepare the application papers, will correct documentation to satisfy local standards, and when required, prepare a translation. These costs are, therefore, approximations and the test application is assumed to cover 22 pages and have 20 total claims with two independent claims (see Table 35).

Not all ''technologically capable'' countries are shown, but if only 30 countries were to be covered by just one patent, a reasonable sum of money is involved. The alternative of doing nothing very often carries a considerable risk.

The cost of a U.S. application will vary depending upon complexity and the hourly billing rate of the attorney or firm. Industry-wide surveys of corporate attorneys as well as those involved with private law firms indicate that 35 to 45 hr may be a reasonable estimate for average time required to prepare an original patent application. If the hourly billing rate is taken as $150, original patent application cost (preparation) can be estimated at $5200 to $6800. If a "typical" application contains 20 total claims and 3 independent claims, filing fees for a patent in the U.S. are $340.

III. INSTRUMENTATION

Instrumentation and control, as a subject, should be integrated with computer control, as discussed below. Advances in understanding fermentation operations and in process optimization depend upon rapid and meaningful measurement of critical parameters. There are many measured variables which can impact fermentation performance (see Table 3, Chapter 1). Not all are amenable to real time measurement and response, nor are all necessary for process optimization. The degree of manipulation of the bioreactor is somewhat more constrained. Measurement can be used to manipulate these responses or variables:

- pH can be controlled by addition of acid/base/substrate.
- Temperature can be controlled by cooling water flow or heat input.
- Growth rate can be controlled by addition of growth stimulants, oxygen transfer rate, temperature, or pH change.
- Addition of inducers or precursors can direct synthesis to desired compounds.
- Foaming tendency is controlled by antifoam addition.
- Oxygen transfer is controlled by impeller power, air input, oxygen supplementation, or substrate control.
- Substrate content is controlled by substrate feed rate.

The objective of process control in fermentation may be considered the rational alteration of environmental conditions to positively impact physiologic activity. While fermentation processes are as amenable to process control as any industrial operation, there are certain unique elements which should be considered in instrumentation and control by any means. Measurements at the cellular (or subcellular) level are difficult or impossible. Fermentation measurements are noisy. Control variables are most often interactive. While a usual chemical catalyst has a level (or slowly changing) activity, enzymatic activity changes rapidly in the course of operation. Many optima (as temperature, pH, DO_2) are not constant in the course of a batch; different optima exist at different physiologic ages. Cell growth rate, enzyme formation rate, and enzyme activity not only vary with time, but are impacted by substrate (reactant) concentration, trace element levels, contaminant (both viable and nonviable) content, product concentration, as well as process parameters. Measurements (*in situ*) must involve establishing and maintenance of sterility. Degradation or reversion of the producing microorganism is always a possibility. Small changes in inputs may have a greatly magnified effect on one or more outputs. Even one of a host of possible upsets may have a catastrophic effect on process outcome. This discussion is not meant as an excuse for poor or nonexistent control strategies. The point being made is that careful attention to detail is essential in planning fermentation process instrumentation (selection, installation, calibration) and control strategies. A good deal of instrumentation and fermentation design emphasis is covered in a book by Solomons.[26] A chapter in a more recently published book covers fermentor instrumentation and control and an introduction to computer-controlled fermentation systems.[166]

A. Temperature

Temperature can be monitored by various means. Normal sterilization range is 121 to 150°C and operating temperature range is 20 to 40°C. It is usually a good idea to use a separate thermistor or other resistance temperature detector selected for each of the above ranges. A "typical" location is usually sufficient for operating temperature control; multiple locations are suggested for measurement of sterilizing conditions. In production fermentors, various problems could result in improper dry sterilization (or sterilization of the clean, empty fermentor prior to charging). Localized pockets of liquid or trapped air might not reach sterilizing temperature in the heating cycle. Appropriately located probes are useful in preventing such occurrences.

Proportional control can be used at production scale with smaller vessels adequately controlled with an on-off system. Calorimetry is possible with a water flowmeter and multiple temperature sensors in place. Normally, one such installation on a single production fermentor is sufficient for introduction of a new process or monitoring of an existing one. A knowledge of peak and average heat loads in fermentation can aid in proper scheduling as well as utility costing for segments of the plant. As is true for every sensor, there must be confidence that the measurement is not only correct for the site, but that the measurement is representative for the volume or region under analytical review. For temperature, a reading of 145°C may be correct for the flow region of the probe; however, a temperature *profile* in the flow regime means sterility was not attained.

Flow microcalorimetry is a means by which minute temperature changes due to metabolic heat development are detected and correlated with biomass. Strictly speaking, this is one means of indirectly measuring biomass. It is a real-time procedure that uses a very small broth volume and does not involve internal probes or other measuring devices. Problems that are encountered with this apparatus are oxygen depletion and pluggage of the flow apparatus. Such a device is external to the fermentor and involves a very small volume of sample. Total heat load of the fermentor is determined by appropriate temperature sensors (temperature of batch, of cooling water input to jacket and/or coils, and exiting water temperature from jacket and/or coils) and flow measurement. The plot of refrigeration requirement vs. time can be used to determine average and peak requirements for the entire fermentation plant. The cooling requirement is a measure of overall metabolic activity plus any other heat inputs to the fermentor. Other heat load inputs are agitator power, air input (power and temperature), any elevated temperature feed streams, and, on occasion, ambient conditions.

B. pH Control

Measurement of pH is very important in biological systems. Many enzymatic reactions exhibit distinct rate optima in a narrow pH band. The progress of cell growth and product formation can often be monitored by pH change. Even if control of pH were not required, process monitoring is desirable and often essential. Since pH optima for growth and product formation may be different, various schemes have been developed to change or maintain system pH. Control of substrate addition rate is one method. Addition of acids or bases (which may or may not be metabolized) is another. Use of controlled gas feeds (especially CO_2 content) is yet a third. Buffer systems can be added initially or in the course of the fermentation, with or without added nutrients.

Measurement of pH is by means of a glass electrode. A glass membrane at the tip of the electrode is sensitive to hydrogen ion activity. Differences in hydrogen ion activity externally (broth sample) and internally (reference solution within electrode) give a potential difference which is measured and magnified. Fairly rugged and steam sterilizable combination electrodes are in general use in fermentation applications. On-off control is usually sufficient if control valve sizing is proper. It may be a good idea to have an alarm system on the valve as well as high-low alarms on the pH amplifier. A timer might monitor interval the control valve is open; an alarm would be initiated if control were not achieved in a preset time. Alternatively (or concurrently), a certain number of valve actuations in a preset interval with control not attained could initiate an alarm. Criticality of pH will dictate whether one or two electrodes are routinely used. If failure of an electrode occurs, manual sampling and manual addition of substrate or neutralizing agent may be adequate to maintain pH in an acceptable range. If a pH band is not suitable for cell growth or reasonable product formation, a second, stand-by electrode must be installed. It is important that the stand-by unit be checked during each run and not simply activated when the primary sensor fails. Another alternative is to have a portable pH flowcell (readily connected to the sample port) which can be installed upon failure of the primary element. This unit can be calibrated in place and an intermittent or continuous flow of broth used to both monitor pH and take samples.

C. Air Flow and Pressure

Air velocity may be measured by any number of convenient and rugged devices. Airflow is usually determined prior to filter sterilization of the input air; this avoids sterility problems and the need to sterilize the measuring system. Velocity can be measured with Pitot tubes, anemometers, or turbine meters, but the usual installation uses a head meter with a sharp-edged orifice. Precautions must be taken on installation (position of orifice plate and its dimensions, pressure taps and their location and dimension) and instrument suppliers give all requirements. A differential pressure (d/p) cell is installed and calibrated so that volumetric flow is expressed. There are well-known charts for critical factors (coefficient of discharge, expansion factor) which are dependent on orifice type and physical properties of the flowing stream. Correction for temperature may be necessary depending upon design criteria. The control element is usually a control valve sized so that usual operation is neither almost closed nor essentially fully opened. In theory, some sort of rotameter or mass flowmeter is just as suitable; however, these are found less frequently at a full production scale.

Pressure in fermentors is measured by a diaphragm gauge so mounted as to prevent solids deposition. The diaphragm should be easily cleanable and steam sterilizable. If standard pressure gauges are used, it is probably a good idea to utilize a flexible diaphragm device to separate process flow from the measuring element. Electrical devices (strain gauges) are far less common for pressure sensing in fermentation practice.

D. RPM and Power Input

There are many convenient means for measuring impeller RPM. A tachometer may be used. Optical (noncontact) devices can be used with some marker or element on the moving shaft. A gear can be mounted internally and connected to some measuring device (electrical or mechanical) which will transmit a signal related to RPM. This is usually one of the simpler fermentation measurements. It is a critical variable for process optimization in many instances. In cell culture or in development of shear-sensitive organisms, careful attention to design and monitoring is required.

Power input to the broth is an important variable. Generally, simpler if less accurate techniques are used. Electrical power (correct for base loading, power factor, agitator seal losses) is often used, but this is indirect. Mathematical correlations, a combination of theoretical and empirical, are available; this is even more indirect. Since power input is related to liquid turnover, oxygen and carbon dioxide mass transfer, homogeneity, temperature control (remember that there is a heat input related to mechanical power input), and maintenance of dissolved oxygen level, it would be a useful parameter to know. A strain gauge dynamometer connected to the shaft within the fermentor below the agitator shaft seal (or above the seal on a bottom entry shaft) will give a direct measurement of shaft power. Such a device is costly to install and difficult to maintain. It would be invaluable in vessels used for scale-up.

Whatever method employed for power measurement, it is prudent to have an alarm which indicates an overload and which will shut down the drive motor. Very often, an interlock between airflow and agitator power is used to insure that the drive motor will not start in an ungassed state. It is wasteful to have an installed drive system selected for an ungassed state when it is never used for liquid mixing only (this presumes continuous sterilization of medium). Gassed power input is appreciably less than ungassed power input for the same liquid volume. Even a relatively low gas input causes a significant reduction in liquid-gas density. Installations are known where two-speed or variable speed drives are used to maximize efficiency and mass transfer. Lower RPM values are used for sterilization or harvest, while higher values are used during aeration. Ideally, agitator power input and gas throughput should be continually varied to minimize cost while maintaining some parameter (as dissolved oxygen level or redox potential) at a desired value. A discussion of scale-up, including maintenance of constant volumetric O_2 transfer rate ($k_L a$), is given a chapter in a text derived from a course in biochemical engineering.[167]

E. Liquid Level

Nonsterile fluid levels can be measured by a number of convenient means. Visual devices are the simplest. A dipstick, float, or gauge glass should not be disdained merely because they have been in use for centuries and are simple. Automatic indication for some control purpose presents certain complexities for these devices. Where some control action is desired or an alarm is to be actuated, float-actuated devices can be used. Physical displacement can be sensed easily.

For level sensing of a sterile fluid (or a liquid that must be totally isolated), indirect means are preferred. For a fermentor, a differential pressure device has proven suitable. A flush-mounted diaphragm sensor can be inserted into the lowest portion of the straight side of a fermentor (through a jacket if necessary).

Output from this pressure sensor is fed to a differential monitor with another pressure input coming from the head pressure sensor usually located on the gas effluent line of the fermentor. Once head pressure is "subtracted", the differential pressure is proportional to the liquid head in the fermentor. One must correct for the volume not "read" by the sensor (essentially volume in bottom dish plus liquid to give initial indication on lower diaphragm). Such a system will *not* correct for air incorporation in the broth (or changes in broth density) so it does not truly measure level unless density is accurately known. Magnetically coupled devices may be used; sterility can be maintained since only a float contacts internal fluid. A follower magnet (inside a completely isolated column) moves along with the float and level is monitored from the follower. The tube that guides the float should be of nonmagnetic material. Capacitance or electrical measurements should be suitable, but because of coating and foaming problems, they have limited use. Radiation devices can also be used, but again, changes in broth density present problems.

Strain gauge systems can be mounted at vessel legs or support plates. Dimensional changes in the gauge (caused by pressure due to supported structure and liquid) results in changes in electrical resistance. This can be monitored. Agitated vessels can be so supported, but vibration and torsion must be compensated for or dampened. Liquid (or mass) measurement is very important in fed batch operations (intermittent withdrawal and recharge) or in continuous fermentation.

F. Liquid Flow

In theory liquid flow can be measured by any differential pressure device or by any one of a number of rate meters. Positive displacement, turbine devices, and mass meters may be used. Vortex-shedding and ultrasonic meters are now available as are radiation devices. For nonsterile application, any device which can handle the fluid, that fulfills cleanliness requirements, and that meets accuracy requirements is suitable. Considerations that enter the selection process involve first cost, installation cost, compatability with control devices and schemes, personnel experience factors, and historical data on the device or similar ones.

Sterile fluid flow is metered by "noncontact" devices, if the sterile stream itself is to be monitored. One example is intermittent or continuous feed (or harvest) from the fermentor during a run. Magnetic flowmeters have no moving parts and calibration or repair can be affected without flow interruption. The sensing tube can be lined with a steam-sterilizable material and the sensing element can be sterilized along with the connecting pipes. Obviously, the flowing fluid must elicit a response from the sensor; this may be a problem for certain nonaqueous systems. Other devices use other indirect means while preventing potential contamination problems due to intruding shafts or moving parts. In a vortex-shedding device, turbulent eddies are created by having some fixed object or vane in the fluid path. Rate of vortex formation and shedding is directly proportional to volumetric flowrate. At sufficiently high Reynolds number, claimed accuracy is described as maintained regardless of other properties such as fluid density, temperature, viscosity, pressure, and temperature. Ultrasonic flowmeters depend upon generation-reception of sound waves and their time delays as flowrate varies.

For small liquid flow, sterilizable head diaphragm pumps may be used. Positive displacement may be varied so that some variation in feed rate is achievable. Liquids with particulates cannot be fed in this manner, but reasonably high pressure heads can be overcome with a diaphragm pump. Another alternative is a peristaltic pump. In such a unit, rollers, disks, or metal "fingers" compress a flexible hose (which can be sterilizable) that results in fluid transport against a pressure gradient. Differential pressures cannot be too high. The pump parts do not contact the fluid flowing within the flexible tubing. Different liquids may be added at different times. Volumetric flow is limited by tubing diameter. Variation in flow occurs as the tubing wears. Simplicity in use often overcomes design drawbacks.

Another simple method for adding fixed volumes of liquid employs an in-line tank. Properly valved to maintain sterility and to permit a pressure difference, a small vessel (or a surge volume in the line) can be filled and emptied one or many times during a run by opening and closing appropriate valves. A weigh tank connected to the fermentor offers a more complex way of monitoring and feeding liquid.

G. Foam Control

Foaming exists in both aerobic and anaerobic fermentations. Foaming is commonly a problem in the former case. Loss of biomass and/or product is undesirable, but potential for contamination or negative environmental impact poses even more severe hazards. Further, a mixture of froth and air contacting the back pressure control valve can cause airflow disturbances. For these reasons, most fermentors have some device for indication of a foam head beyond an "acceptable" level. Conductance or impedance systems can be used. One or more electrodes are inserted into the vessel head (or less frequently, in the air exit line) and insulated from the vessel shell. A variable sensitivity adjustment is incorporated into the electrical circuitry so that a threshold conductivity must be exceeded before the control system is activated. Many fermentor agitators have a mechanical foam breaker (paddle, rods, screen, disk, or a combination) attached to the shaft well above the unaerated liquid height. The foam detector probe will usually be contacted after the mechanical device can no longer cope with the foam head. High speed devices have been designed with separate and independent drives. Location is at, near, or in the air exhaust line (which may be 10-in.-diameter pipe). These high speed machines depend upon high centrifugal forces to separate liquid and gas. Sonic devices have also been employed. Energy input systems usually operate continuously (or continuously after a certain point in the cycle) and are not normally connected to the foam probe(s).

Once the foam probe is continually contacted, some control sequence results in introduction (as a spray or jet stream, as an aerosol in the air input, or as a liquid above or below the broth surface) of a defoaming agent. There are a number of timers (variable settings) in the control circuit:

1. Delay: the probe should be engaged for some continuous interval before sequencing begins; otherwise, splashing would cause defoamer addition.
2. Shot length: interval of valve opening should be set and variation is desirable; early in cycle, short shot length is suitable, while a longer interval might be more efficient later in cycle.
3. Interval between shots: some period is desirable between defoamer valve openings, both to allow for a time lag in defoamer efficacy and to prevent overaddition.
4. Time for achieving control: a variable timer (start at probe contact) which resets when control achieved should indicate an alarm if control not achieved in a preset interval.

Typical ranges and settings for timers are listed below. It should be understood that many variables are involved in selecting timer range and setting; final determination is empirical rather than derived from theoretical considerations.

Defoamer timer	Range	Usual setting
Delay	0—60 sec	5—10 sec
Addition (shot length)	0—60 sec	5—10 sec
Interval	0—10 min	30—60 sec
No control (alarm)	0—15 min	2—4 min
Manual override (shot)	0—60 sec	5—10 sec

Manual override should be provided along with an alarm for some number of sequential valve openings without affecting control. If volumetric flow of defoamer is not provided, an integrating device or cumulative timer should be installed so there is an estimate of defoamer used per batch. The ideal defoamer will not only be effective in the gas-liquid system, but will also be easy to sterilize, be inexpensive, and have no impact on microbial growth product formation or later isolation. The ideal defoamer has not been found, so a reasonable number of pilot tests are needed to minimize negative impacts. A defoamer control device can also operate from changes in power input to the fermentor drive. When the mechanical foam breaker is contacted by foam, power draw rises. Total electrical power draw (or current drawn) can be monitored and when surges are noted or a continuous increase occurs, defoamer addition can be initiated. Another useful technique is to incorporate some level of defoamer in medium make-up or in nutrient additions in the course of the run.

Whether added with nutrient feed or separately, defoamer fluid (which may not be readily miscible with water) must be sterilized before it is fed. A batch feed tank (one of two or three, charged and sterilized in sequence) can be used to feed one fermentor or multiple units through a common header. A recirculating system (sterile vessel and sterile headers) can also be used with defoamer maintained at elevated temperatures if the material is stable at those conditions.

H. Effluent Gas Analysis

Most fermentation plants have continuous on-line capability for off-gas analysis for O_2 and CO_2. If a dedicated unit is not installed, there may be a research or pilot plant unit that can be used intermittently to monitor plant vessels. While it is a relatively simple matter to monitor one fermentor, monitoring one or two dozen vessels is somewhat more complex.

In either case, appropriate pressure reduction, safety devices (prevention of foam or liquid entry to analyzers), knowledge of lag time, and humidity control should be accomplished. For multiple vessel use, a multiplexing system is needed. Samples from all fermentors should flow continuously (minimize lag) and are vented while the selected sample is monitored. There should be provision for manual override to select any fermentor, capability for continuous monitoring of one unit (override selector), capability for by-passing any one or a combination of inputs, and capability for manual or automatic calibration with standard gas of known composition. For a dozen vessels with a 2-min sample cycle, each fermentor will be checked roughly twice per hour. This may or may not be sufficient; time in overall cycle and type of fermentation will determine whether data input is adequate. In general, however, such a frequency will be adequate once acceptable ranges for the process are known.

Standard units use paramagnetic analyzers for O_2 and IR for CO_2. Gas chromatographic methods and mass spectrometry may be used as well.

I. Mass Spectrometry

Mass spectrometers have been coupled to fermentor effluent gases. Advantages of such analysis are extremely fast response and capability to analyze a number of components. An industrial mass spectrometer should be able to analyze five compounds in gas streams at the rate of four samples per minute.[168] The advantage of coupling such a device to a computer for real-time decision making is obvious.

Analytical power can be enhanced by a series connection of gas chromatographic separation devices. Allied techniques such as nuclear magnetic resonance, infrared spectroscopy, and flow cytometry, have been tested in laboratory environments.

While the historical means to monitor off-gas continue in use (infrared analysis for CO_2 and paramagnetic analysis for O_2), mass spectrometry has been applied to effluent gas analysis. Use of the Perkin Elmer MGA-1200 is described in a recent article.[169]

The drawbacks are well known — initial cost may be somewhat over $50,000 and maintenance and repair are both complicated and expensive. On the positive side, reliability is said to be excellent and there is little to no calibration drift. Examples are given in the article for different fermentations, the distributed digital control system is described, and sample calculations are detailed. Other units, such as the DAI Model 800 GC/MS mass spec, cost from $55,000 to $75,000, depending upon attachments. These can be considered at the low end of the price spectrum. Kratos Analytical offers a unit (MS 25 RFA) that is capable of GC/LC and SFC (supercritical fluid chromatography). Such a unit can be used for a great deal more than rapid gas and volatiles analysis; characterization of complex bioorganic molecules is possible. Cost is approximately $200,000 (costs are early 1985). State-of-the-art units from Finnigan or VG Instruments cost from $250,000 to $350,000.

J. Dissolved Oxygen

Dissolved oxygen (DO_2) is measured by special membrane electrodes which are sterilizable in-place. DO_2 is a function of oxygen uptake rate of the culture, inlet O_2 concentration, airflow rate, impeller power input, vessel pressure, vessel geometry, and presence of surface-active agents. There may be other subtle factors involved, but to make some impact on DO_2, there are only a limited number of practical things to do:

1. Change impeller power input
2. Limit substrate addition rate
3. Change O_2 level in incoming air
4. Change airflow rate
5. Change vessel pressure

In a laboratory or pilot fermentor, positioning of the probe does not present problems. In a production vessel, DO_2 may show appreciable variation with probe location. The variation occurs in both axial and radial directions. Ideally, multiple probes should be used. This is both costly and cumbersome. It is probably a better idea to establish a range of DO_2 content with time for a "good" fermentation process result and other for a "poor" result. This sort of effort presumes DO_2 content is a critical variable and other process upsets are neither more significant nor more noisy. It may be possible to develop a critical DO_2 level for one, fixed location in the production vessel. Even if there are wide variations elsewhere in the broth, one still has a "control point" and monitor for projecting performance. The DO_2 response may be used, in the simplest sense, to program airflow. If a dual or variable speed motor is in place, two input variables can be changed. While a microprocessor can be used, a multivessel installation can be controlled with a central computer. Further, an optimum strategy to control DO_2 can be developed, since plant-wide inputs can be entered and logged. Should the total plant electrical demand be reaching a new peak, increasing motor horsepower would not be a desired response. Switching on a stand-by compressor to increase airflow would not be automatic, but would depend upon overall plant demands, cost of energy, and other programmed constraints. At the very least, alternative strategies and costs can be rapidly developed by the computer so that a more reasoned human intervention occurs.

Carbon dioxide electrodes have been developed and are available; however, industrial use seems limited. The CO_2 probe operates by detecting a pH change related to CO_2 permeation across a membrane. It is also possible to compute pCO_2 from considerations of O_2 and CO_2 mass transfer rates, flow rate, pressure, and Henry's law constant. It appears, however, that measurement of exit gas CO_2 concentration is usually sufficient for control purposes. Some fermentation processes show inhibitory effects at elevated CO_2 levels. This is true even if oxygen transfer is adequate. Monitoring exit CO_2 level and controlling airflow normally provides a sufficient safety margin.

Both dissolved O_2 and dissolved CO_2 can be measured by means of a coil of porous Teflon® (or other material) immersed in the fermentation medium. A carrier gas is passed through the coil at a carefully controlled rate. Diffusion of O_2 and/or CO_2 (as well as other dissolved gases) through the membrane occurs. The amount will depend on the concentration driving force, area of the coil immersed, and rate of carrier gas flow. The sensing devices must be sensitive to rather low concentrations of gases and will, therefore, be more costly. A mass spectrometer would be suitable and, under appropriate conditions, less complex sensors can be used.

K. Redox Potential

Redox potential is a somewhat more vague (that is, nonspecific) measurement. Many chemical entities and their states in the fermentor broth contribute to redox. The measurement is usually made with a platinum electrode in conjunction with a calomel or Ag/AgCl reference electrode. It is possible to use semipermeable or impermeable membranes or coatings to prevent oxygen (or some other compound or radical) from affecting the reading. This is done because any compound capable of oxidation or reduction will give an electrode response. Redox potential can be negative or positive; therefore, there is a wide spectrum of fermentor broth "states" that can be monitored. In tissue culture especially, there appears to be rather limited redox optima. Therefore, changes in media or in gas perfusion can be made to reach or maintain desired redox readings. Redox readings prove useful in anaerobic fermentations (DO_2 levels will be zero or relatively meaningless).

There has been a surge of interest in control at the cellular level. Concurrent with the postulated need is the question of whether such physiological monitoring and control are truly required. Where manufacturing cost is a very low percentage of market value, optimum control may not offer an economic incentive.

Still, developments in mammalial cell culture, plant tissue culture, and the synthesis of esoteric microbial products do present process control problems which may not be amenable to classical monitoring/control networks. A review of biosensors — state of the art and needs — is available.[170] The major needs appear to be

- Measurement of microbial biomass
- Reliable carbon and nitrogen sensors
- Sensors for selected amino acids
- Sensors for selected proteins and enzymes

There are many problems to be overcome even before essential ruggedness is built into industrial monitors. Methods development on a laboratory scale is needed. Fouling, drift, poisoning, loss of sensitivity, and premature failure are a few of the problems encountered. The need for establishment and maintenance of sterility is obvious, if *in situ* probes are the goal. However, if an effluent stream (either sample or product overflow) can be used, sterility requirement may be relaxed. Reports have been made utilizing enzyme-sensitized amperometric or potentiometric electrodes, but once again, application to a wide range of biological systems is limited and extension of utility is problematic. It is likely that until a definitive benefit is shown and a clearer need exists, the biosensors described will remain laboratory tools.

Enzyme electrodes have found numerous applications in analytical work. Carbohydrates of various sorts and amino acids have been assayed in a convenient manner by means of immobilized enzyme probes. In these cases, standardization is relatively simple as is cleaning; further, there is no need for sterilization. Turner describes procedures for making, and methods of utilizing, modified biosensor electrodes having immobilized mediators.[171] The operation involved flow cell design with microprocessor control. Sterility was maintained and drift compensation and automatic recalibration were achieved. Modified electrodes with immobilized mediators shuttle electrons have been the biologic and inorganic components of the system. He has selected ferrocene and some of its derivatives, since they meet many of the operational requirements of such a material:

- Ability to undergo rapid redox reactions in a practical electrode and a biological system
- Stability
- Lack of side reactions during electron transfer
- Redox potential sufficiently different from other electrochemically active components
- Consistency over a wide and relevant pH range
- Suitable for immobilization or incorporation
- Nontoxicity

Performance of ferrocene-mediated enzyme electrodes (various sensors, as amino acids, CO, sugars, lactate, methanol, NADH) are detailed as is construction of the electrodes and measuring circuits.

Intracellular activity can be monitored by an existing apparatus. NAD/NADH ratio can be measured on-line using a fluorometer and specially designed probe. Various types of equipment are available for measuring ATP concentration using luciferase; this is accomplished off-line.

One commercial probe (Fluro Measure™ System by Bio Chem Technology) uses a surface-type fluorometer within a probe that is insertable into a fermentor through a special 25-mm fitting and is designed for *in situ* steam sterilization. Culture fluorescence measurements have been related to cell concentration, growth rate, specific growth rate, sugar use rate, diphasic transition, oxygen, and other nutrient limitations. The detector, transmitter, and connector (for transmitting 4 to 6 mA signal outputs — NAD[H] and lamp intensity) cost $10,550 (all prices mid-1985). Associated computer interface package with data acquisition/ analysis software for various IBM computers adds approximately $5000. An expanded system (three detectors) with enhanced interface package and enhanced data acquisition/analysis package plus an IBM-AT™ is available for $45,000. Various rental arrangements can also be made to evaluate the equipment.

Other rather more complex apparatus are used to measure and analyze cells. A Coulter counter can perform cell counting and cell sizing. Presence of particles or dead cells would cause interference. A laser flow microfluorometer can measure the distribution function of macromolecular contents and cell sizes. A fluorescent dye is used to stain cells. A variety of cell constituents (proteins, enzymes, RNA, DNA) can be measured. This equipment is very complex and costly.

Biosensors can also be employed to monitor bioproducts and foods for microbial contamination, toxins, and very low levels of specific chemical components.[172] DNA probes and monoclonal antibodies are used as diagnostic tools for specific identification of microorganisms. Monoclonals can also be used as test reagents for detection of specific compounds in food, as well as for specific protein related to many different physiological states. Nucleic acid sequences can be used as probes in hybridization assays for many different microorganisms. Viruses, bacteria, protozoans, etc. can be detected in tissue, physiological fluids, water (such as effluent streams), as well as in foods. Commercial test kits give very rapid and reproducible results; classical methods, aside from being complex, often took many days or weeks. After identifying a probe sequence, a label or reporter molecule is selected. The reporter molecule (examples are radioisotope, biotin, hapten, or specific enzymes) must then be incorporated into the probe sequence and an appropriate analytical method devised for detection of the probe sequence. Monoclonal-based immunoassays for *Salmonella* in food are commercially available. If a monoclonal antibody can be prepared (presuming some antigenic response is elicited), one can, in theory, prepare an assay for minute quantities of that antigen. Assay procedures for the desired fermentation product (as well as by-products) can be developed for monitoring the purification process as well as controlling discharge (or minimizing loss) of selected compounds during processing.

Most instrumentation for fermentation applications is scale independent; i.e., with appropriate holders, leads, and mechanical support, a pH system, for example, could be used at any scale of operation. Certainly, a measurement and control system for a 500-gal fermentor is generally suitable for a vessel 10 or 100 times the size. Installation and remote monitoring would add costs to the basic unit.

Measurement of pH is about as standard as any fermentor sensor is today. A host of sterilizable electrodes is available. Cost of sensor and amplifier is about $1000 to $1500. One supplier of pilot plant equipment has a pH control module plus addition pump plus timers for dosing times (acid and base) for approximately $1400. The same company supplies these additional items (all prices are early 1986):

	Price in dollars
Foam control monitor/dosing with pump	750
Dissolved oxygen measurement (electrode/amplifier)	2300
O_2 control system (flow or agitator speed) with special valves	1700
Redox measurement (electrode/meter)	1200
Six-color recorder	3400

The measurements and controls which might be applied to a pilot fermentor are shown in Figure 19. All these controls could be installed in production fermentors as well; normally, economic constraints limit the number of instruments to those absolutely essential to process monitoring and control. It is prudent to select indication and/or recording and/or control with/without alarms for each variable. This not only saves on installation cost, but also saves on maintenance, records retention, and reduced operator error. In a multifermentor plant, it may be worthwhile to have one very highly instrumented vessel at the production scale. With appropriate flexibility in design for this one vessel, the additional expense incurred may be returned many times over by (1) more rapid scale-up, (2) process optimization, and (3) energy use reduction after careful use of such an "experimental" unit.

Cost of instrumentation and control is given in Table 36. Costs are approximate. A distributed control system (assume 50 loops) is shown for comparison. The hardware for such a distributed control system would cost $60,000 to $70,000 with software costs estimated at 30 to 40%, in addition. Operator display would be $25,000 per console. For pneumatic or single loop microprocessor, a control panel would cost $1500 to $2000 per lineal foot. Installation cost (as percent of total equipment cost) would be 25 to 35% for pneumatic and approximately 30% for microprocessor or distributed control. Speed (RPM) control is estimated for a 500-HP motor and drive.

IV. COMPUTER CONTROL

As with other complex industrial processes, fermentation operations are amenable to computer applications. While initial installations emphasized control, a shift toward process optimization has occurred. Modeling and optimization are of interest to industry insofar as economic justification can be determined and achieved. Computer control and monitoring offer immediate advantages compared to conventional analog systems:

1. Repetitive, step-wise processes can be programmed and will proceed only upon successful completion of prior steps. A prime example is sterilization.
2. Complex algorithms can be programmed and readily modified.
3. Data collection, in a timely fashion, is not dependent upon human intervention. This is a great advantage when many fermentors are operated with differing processes. Plots can be readily generated and statistical analyses performed.

Once these advantages were realized, material and energy balances could be determined and optimization routines developed. The simple cases involve cell mass, as in brewers' yeast. As was noted in a review article, more involved control objectives require an evolution of more complex models and control strategies.[173] Some attempts have been made to determine key process parameters for higher productivities while determining costs for the desired inputs. If differing agitation-aeration inputs are known to result in various annual product outputs, an economic balance can be calculated based upon the cost of energy (to motors and drives of mixers and compressors while compensating for changes in energy efficiency due to load factors) to affect each output. Minimum cost per unit of product will probably not occur at peak output per month or year. If sales margin is very high and demand is high, such niceties are not needed. More product, even at above minimum cost, is the goal. Where inventory is sufficient and margins are more moderate, optimization routines can have a significant impact on cash flow.

Initial success in analysis and control of fermentation process can be traced to laboratory and pilot plant applications. A review of simulation strategies applied to microbial systems from the late 1960s to early 1980s is given by Knorre.[174] A computer system (called ZIMET) for fermentors from 3 to 3000 ℓ installed in East Germany is described. The system appears rather conventional with measurement of pH, pO_2, pressure, temperature, RPM, electric

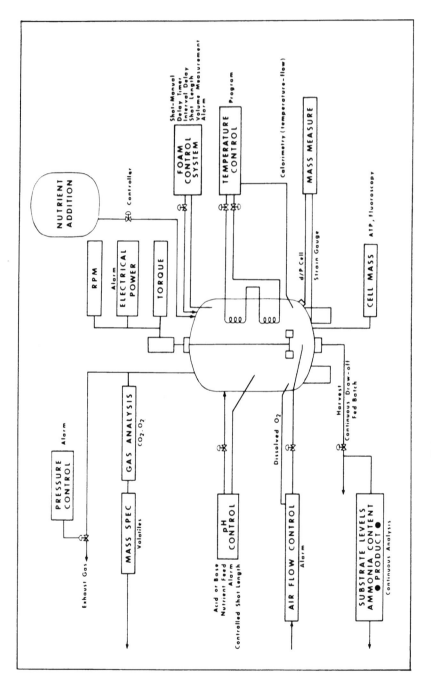

FIGURE 19.   Control potentials in fermentation.

Table 36
COST OF INSTRUMENTATION AND CONTROL[a]

	Pneumatic	Single-loop microprocessor	Distributed
Control, with valve			
Air flow	4.0—4.5	3.8—4.4	1.3—1.8
Pressure	3.5—4.0	3.3—3.8	1.0—1.2
pH	4.0—4.5	3.5—4.0	1.0—1.2
Temp (dual range)	3.5—4.0	3.0—3.5	2.0—2.5
Flow (magmeter)	6.0—6.5	5.5—6.0	4.0—4.5
Control, without valve			
DO_2	2.0—2.5	2.0—2.5	Inclusive
Level	2.0—2.5	2.0—2.5	Inclusive
Defoamer	2.0—2.2	2.0—2.2	Inclusive
Speed (variable frequency drive)	30—40	30—40	30—35
Indicator analyzers			
RPM (per point)	0.3	0.3[b]	Inclusive
Alarm (30 points + annunciator)	6	6[a]	Inclusive
Utility monitor (3 sensors)	6	6[a]	4
Integrator (per point)	1	0.5[b]	Inclusive
O_2	3	3	3
CO_2	20—25	20—25	20—25
AutoAnalyzer	20—25	20—25	20—25
pH meter	1.0—1.5	1.0—1.5	1.0-1.5
Recorder			
Three pen	1.5—2.0	2.0—2.5	—
Multipoint (30)	—	15	—

[a]   Units are in thousands of dollars.
[b]   Indicators are single-loop electronic.

power input, fermentor weight, OD, effluent O_2 and CO_2, and control of pH, pO_2, temperature, and RPM. Off-line subroutines are available for manual and automatic data handling, calculation of derived variables, graphic representation, process model investigations, parameter estimation, and optimal control. Examples are given for analysis and process optimization of the turimycin fermentation. Both on- and off-line schemes were attempted and data are presented for both cases.

Two papers by Meiners[175,176] discuss a computer coupled system with 20 fermentors having volumes between 10 and 5000 ℓ. The output file is described completely. Process variables are subdivided:

Measurements: these are followed by differentials of the measurement functions to show changes of state variables; standard deviations calculated and stored
Manipulating variables: as RPM, air flow, feeds
Objective variables: calculated values such as growth, production, substrate use, yield, etc.

Applications of microprocessors on selected measurement or control functions are also given. The plan of computer coupling of the pilot plant is given. The layout of the analog/digital converter is shown. Data files and program examples are shown as well. The main objectives of the system were data acquisition, simple process control, and off-line analysis. The articles present a hardware approach (meant both literally and figuratively) and present actual performance results after many years of testing and modification. Practical experiences and operational results are well worth the time to review these papers.

Spark discusses an integrated data base in terms of information needed to control and analyze bioprocessing data.[177] Data handling, both to and from a production plant, is clearly important for scale-up and optimization. In an integrated facility (lab, pilot plant, production) information should flow freely to provide scale-up and scale-down relationships. The communication system should allow entry of experimental designs, appropriate operating procedures (SOP) for selected experiments, execution, data collection, storage, and analysis. The system is interactive for process design, equipment operations, and data analysis. Data types are

- Current — present process information
- Historical — process completed, analysis, record keeping
- Future — set-up data, batch cards, process not yet started
- Archived — on computer tapes
- Summaries — various synopses

The practical methods of data introduction and sequencing are described. Formatting is shown. Since all stages of development and production are fitted into a common system, passage of process development information will be simplified. Confusion which accompanies a change of venue (or scale) is avoided. Statistical analyses are also simplified. There is an ongoing, live linkage between researchers and production management through the common information channels.

Another application study is given by Williams.[178] Fermentation of bakers' yeast was studied using an on-line real-time adaptive package. Controlled responses of the fermentor are presented as a next step in solution of the control problem. The computer control loop automatically tunes direct digital control feedback loops without significant manual intervention. Parameter estimation techniques are used; recursive least-squares estimation was selected. The variables used in the adaptive algorithm were CO_2, O_2, alcohol in effluent gas, DO_2 in broth and substrate, and nutrient flow rate. A schematic is given which shows inputs to the fermentor, and measured outputs and computer interfacing. Structure of the adaptive computer controls system is also shown. A mathematical formulation is presented for the dynamic model of yeast fermentation. This is in the form of a set of nonlinear first-order differential equations with a single input control variable. The technique is applied to process control with a single input and multiple output accessible control and state variables. Input control is affected by substrate (sugar) feed rate. The adaptive control package was able to maintain demand level trajectory for four parameters (DO_2, CO_2, alcohol, respiratory quotient). Multivariable input control (RPM and nutrient feed rate) was successfully implemented. One conclusion is that the fermentation system can be accurately controlled, but optimization requires careful selection of control parameters and the performance criterion. This is not a surprising result. Another, more surprising projection is that adaptive control could eliminate detailed understanding of fermentation and similar processes while achieving some biological "target". One would need a reasonable element of good fortune to achieve such a result, since selection of input variables and output responses would require a reasonable knowledge of the process. If the authors mean that every reaction step in a long synthesis need not be known or kinetics need not be well defined, then optimization can indeed be achieved without such detail. In fact, that has been the historic performance of fermentation processes over the past 40 years.

Another paper, while describing a food plant, describes instrumentation and techniques which are of major import to the fermentation industry.[179] Sterilization is an obvious requirement for certain processed foods and is equally critical for most fermentations. A system (or actually three duplicate systems) is described to automate the complete production cycle of 24 batch cookers. The objective is clearly stated and its applicability to fermentation processing should be obvious. "The object of the commercial production application was to automate the process cycle control of 24 retorts, improve the control accuracy at strategic points to ensure a higher degree of repeatability between batch processes, improve the accuracy and level of process documentation, and to improve general process efficiency in terms of energy management and operator efficiency." (p. 360) Plant hardware is detailed and microcomputer architecture is described. One relatively small section of the article discusses the human interface. It is worth quoting a relevant section, "The process plant operators are usually manually skilled people with a significant amount of practical experience of the behavior of both the plant and the product. They are usually the people who have the largest influence on batch production rates and product quality. Only by incorporating some of their distilled wisdoms within the logic of control software will the control systems be capable of producing even part of a skilled operator's effect on the process plant." (p. 364) Many designs overlook this key point. It should be emphasized and reemphasized all during planning and introduction of any new control scheme.

A book can also be consulted for recent developments in process and computer control.[180] There is a useful tabulation of control system design considerations that should be reviewed before architecture is even considered:

- Current or proposed standard plant operating procedures
- Automation level
- Process management level
- Personnel requirement
- Reliability
- Security needs
- Control performance criteria
- Project scope (including time frame)
- Operator job description
- Accuracy required
- System responsibility
- Process dynamics and economics

The chapter on process control reviews needs in a fermentation plant. Simple feedback control is a major task in such a plant. There may be 200 or more feedback controllers and perhaps half a dozen advanced control loops. Any advanced control application (especially one involving optimization) requires precise and accurate operation of basic feedback loops. The review covers noncomputer-based systems, computer-based systems, and those involving distributed and hierarchical concepts. The various potentials are listed:

Reduced time requirements	Optimization routines
Greater accuracy	Inventory control
Sterilization (consistency)	Diagnostic analysis
Energy reduction	DO_2 optimization
Simplified changeover	Reproducibility
Data logs (regulatory)	

Three "must" considerations for successful implementation of a control scheme are

1.   Firm base of single loop control established before higher level automation is considered
2.   Knowledge of the process required before data can be used
3.   Long-term control strategy needed — implications of the control system should be understood

There is a useful glossary of some 300 terms used in computer applications.

What is apparent is that microprocessor technology is now generally applied to fermentation installations, as is the case generally in the process industries. Flexibility and reliability of the systems have been enhanced. Networks involving operator interfaces, video color displays, high speed communications, historical data logging, and analysis are common in fermentation plants. Novel control strategies that vary with time in cycle can be coupled with control enforcement (required operator action and interface) in a simple manner. Changes in control strategy can be made readily.

One company in the U.S. is selling a "multiple fermentor computer system" or MFCS developed by Rintekno of Finland (Xertex Corp., Santa Fe Springs, Calif.). The base system consists of both hardware and software for multiunit process control and data acquisition. The total system can monitor and/or control up to 50 fermentors; operator interface is supplied by means of full graphic color terminals, a special keyboard, and report and screen printers. As one would expect, these functions are performed:

1.   Supervisory — operates and controls, optimizes control strategies using algorithms, sequences from recipe library (which is expandable), preparations for upcoming run can be made before ongoing run completed
2.   Controls — check measurements for validity, adjust set points, sequences, experimental replication, charging and sterilizing included, various control algorithms
3.   Documents — standard reports, batch specifications, historical data, log of key events, incidence/duration of alarms, all historical data can be used for modeling, optimization, prediction
4.   Operator interface — color displays, alarm(s), deviations (six variables/fermentor), plots graphs, short/long-term trends, laboratory data entry, trend group displays, process diagrams with current status of equipment, all operator actions which impact process are logged and stored

The standard reports include: batch ticket, batch report, shift report, daily report, alarm state report, alarm history report. System hardware is as follows: Digital Equipment Corp. (DEC) PDP-11 series Q-bus computers using MICRO-RSX system; MFCS process interface is Analog Devices μMAC 5000; software is written in Macro-11 and DEC Fortran 77.

A review of file treatments will show capability of the system. All stages are stored in the "fermentation stage" file; for each stage, specific subroutines are available. Subroutines include information about variables in use, variables under control, setpoints, and alarm limits.

1.   Select recipe and profile(s) from libraries
2.   Initiation (sequencing of washing, charging, sterilization, inoculation, cultivation, termination, harvest)
3.   On-line operation including data acquisition, monitoring, control (to displays, printer, interface)
4.   Off-line operation (laboratory data, off-line calculations)
5.   Long-term data storage

165

MFCS contains a basic set of calculation algorithms for these variables:

OTR	Heat/OUR ratio
OUR	Heat/CO_2 evolution ratio
CO_2 evolution rate	Cell mass/O_2 yield
Heat generation rate	Cell mass/substrate yield
Productivity	Growth rate
RQ	Specific growth rate
Henry's law constant	Heat of evaporation
Saturation value (DO_2)	Heat of aeration
DO_2	Dilution rate

New algorithms can be introduced. Each calculation is preceded by a relevancy check; this effectively screens erroneous input data. System pricing is as follows (early 1986), but the company should be contacted for an up-to-date quote:

	$000
Base system	140
Hardware	
Software	
Start-up	
Interfacing	
Configuration	
Training	
Software options	34
Fortran 77 license	
MFCS source code/documentation	
Hardware options	
Tape drive and controller	3.7
Multiplexer card	1
Alphanumeric video terminal	1.3
Color graphic workstation terminal	5.2
Printer	2.3
Multifunction keyboard	3

System elements: 108 analog inputs, 48 analog outputs (4—20 MADC)
144 digital inputs, 144 digital outputs (115 VAC)

It appears that a relatively complete system can be assembled for a reasonably sized lab and pilot plant (even some production unit capability) for somewhere near $200,000. This is only one element of the computer control package (discussed below for a complete plant).

Optimization is one of the major objectives of computer control. This desired result is directly related to appropriate monitoring devices and manipulation of process responses. Optimization was included in an extensive review of computer applications to fermentation.[173] The authors detail many references for fermentation operations that involve global optimization. The authors note a point that has been stressed here and elsewhere. Optimization should not be limited to fermentor performance only if cost is to be minimized (or profit maximized). Fermentor operations can be optimized for one or more responses; however, subsequent recovery steps are often adversely affected. Resulting *overall* cost is not at a minimum value. An optimal control strategy should include market demand, raw material costs, waste treatment costs, environmental and regulatory impact, as well as fermentation and extraction operations. Computer programs can be established not only for a single process or product, but can be formulated to present an optimal scheduling regime for a multiproduct plant. All steps of the fermentation sequence (pretreatment, sterilization, inoculation, feed sequence, harvest) can be coupled to multifunctional processes with multiple use extraction equipment to give an optimal use profile for varying production levels of the different products. In general, the hierarchical structure is not different from that in the chemical industry at large. The chemical industry is becoming aware of the utility of batch scheduling via linear programming to improve controls and recipe management while making optimum use of existing equipment. Assignment of recipes to equipment (especially to a multifunc-tional extraction operation, in a fermentation plant) that is coordinated to company marketing needs as well as equipment availability would have an economic impact if a proper algorithm were selected. Such a procedure, with examples, is given for numerous charge vessels, reactors, crystallizers, centrifuges, and driers.[181]

Cooney[182] has reviewed several examples of on-line process control as applied to fer-mentation. The review emphasizes optimization of fed-batch cultures; i.e., those in which one or more raw material feed streams are controlled in response to the metabolism of the growing organism. In this manner, a low and relatively constant substrate concentration is maintained. In the ideal fed-batch process, the organism would be insensitive to low substrate concentrations (inhibited by high levels) and also be insensitive to high product concentration. Since the substrate feed rate is a function of growth rate, conversion yield, fermentor volume, and cell content and the feed rate may vary depending upon the interval in the overall cycle, it seems well suited to computer controlled response. Examples given are

- Bakers' yeast on glucose — achieve peak conversion
- Yeast single cell protein on methanol — substrate control
- Heparinase production — low S, controlled glucose culture
- Penicillin production — product optimization with reproducibility

It is noted that a lack of appropriate sensors may limit industrial applicability; however, indirect measurements coupled with an improved understanding of microbial physiology may provide sufficient input to allow useful computer control strategies.

Computer applications to the fermentation industry are not limited to plant operations. Generalized procedures, such as those noted below, can provide access in these areas:

Bibliographic services	Electronic mail
Project management	Laboratory notes
File sharing	Recipes
Databases	Technical protocols
Laboratory analyses	Equipment inventories

Computer programs have been developed which involve genetic engineering applications. A specific system has been designed for support of genetic engineering research at Eli Lilly and Company.[183] The programs developed are used to perform routine calculations, simulate plasmid or sequence construction, nucleic acid and amino acid sequence analysis, as well as evolutionary studies. Samples of entry menus and printouts are given. Since there are expanding databases containing protein sequence information, a variety of sequence database management, searching, and display programs were developed. There are programs that simulate most genetic engineering laboratory manipulation; computer-generated restriction maps can be generated. Storage, retrieval, and display of circular restriction maps have all been achieved. The paper gives an example concerning the cloning of a hypothetical cDNA from a hypothetical gene. More advanced systems are described (so-called "expert" systems) which can perform logical sequencing and can design cloning experiments. A final point concerns the potential for an automated laboratory connected to an expert system. Not only would computer printouts be generated, but actual experimentation could be performed to deconstruct and reconstruct bioactive compounds.

The number of programs in biotechnology software is increasing. As the amount of data proliferates, the only rational approach to storage and retrieval points is use of computer techniques. Computer-aided processes in biotechnology have been summarized:[184]

- Real-time data acquisition
- On-line data collection/correlation
- Data storage and retrieval
- Nucleic acid sequence analysis
- Protein sequence analysis
- Simulation
- Experimental design

Most operations can be performed on personal computers; simulation and design require more complex machines.

Battelle Northwest has a software package that applies CAD (computer aided design) to genetic engineering.[185] The program calls up stored data on genetic profiles, counts the number of DNA building blocks, and finds homologies and maps restriction sites. It can display a whole plasmid (to 4000 bases) or focus on an 8-unit-long section. The software is designed to run on VAX or mini-VAX computers (Digital Equipment) and will cost $20,000 to $40,000.

The burgeoning development in computer use for pharmaceutical processing has led to major changes in government inspection of such operations. Computers are in use in these areas:[186]

- Maintenance of quarantine systems for drug components
- Control of significant steps in manufacture
- Control of laboratory functions
- Management of warehousing and distribution

In the government publication noted above, a complete guide is given to reviewing all elements of the user system. The key points to cover include:

- Location (environment, distances)
- Signal conversion
- Hardware and software validation
- Software development and security
- Development procedures and programming standards

The instructions relate various items in the control scheme to relevant CGMP regulations (with exact reference to the Code of Federal Regulations). Every item which was covered and reviewed under a "manual" system must receive appropriate review and validation under any new system. There may be differences in appearance and approach, but certain requirements are unchanged, regardless of whether electronic media or paper is used for information storage. The instructions (or SOP) relating to software refer to each of these subjects:

- Record controls
- Record access
- Record medium
- Record retention
- Computer programs
- Record review
- QC record review
- Double check on computer
- Documentation
- Reproduction accuracy
- NDA considerations
- Updates and revisions

Systems must be secured against unauthorized entry or change.

Here is a clear example that shows that planning of a computer installation, and not only of a process or packaging line, must include appropriate internal checks and procedures for regulatory review. While drug manufacture is the main subject of the publication noted, there is a clear and obvious extension to GLP and toxicology testing.

Plant-wide computer control offers a number of advantages. There has been a shift to fully integrated hierarchical systems. One mainframe manages the system with mini- or microcomputers in a network with each responsible for some subset or aspect of the total operation. The names of each level of computational complexity may vary, but, in general, the pyramidal structure will start with a corporate or management level mainframe. This can be considered the apex. There is a data link between it and the next level — plant or factory level at one or more locations. Actually, there is a complex linkage (or bus) between various levels with information moving in discrete bundles on a more or less continuous basis. Movement is bidirectional; failure of a subsystem will not affect other operating subsets. The second (plant or factory) level computer may be a mini. This is connected — via a LAN (local area network) — to a third level of PCs or minis which can be considered "area" level. Each product line may be considered an "area". The QC lab and maintenance are other "areas". Each area has ties to the fourth, or line, level. Each line level device would control the fifth level that contains specific sensors, microprocessors, control devices. The final level might be a reactor, a machine, a fermentor, a crystallizer, etc. Interactive processing becomes a clear possibility. Adaptive control becomes a possibility; the entire plant (or part of it) can adjust to changing inputs to maintain throughput or product quality. If there is an operational history that can be programmed (and this has already been accomplished in certain industries) it is possible to develop an "expert" system. The data bank is utilized to store historical responses to many different stimuli (such as upsets). The program can be compared to a "knowledge bank"; responses will occur without human intervention. This can only result if human know-how and experiential knowledge are transformed into such an expert system.

There are many ways that a properly designed local area network might improve fermentation plant operations. Some will have major significance and others will be convenience items. The various potentials are listed below. Some concepts might have defined payouts, while others show advantages that are difficult to quantify in dollars. When a costly pharmaceutical is to be synthesized, extracted, and purified, one should not discount the intangible elements of improved communication and data correlation.

Use of computer control on an industrial scale can offer these advantages to a fermentation plant:

Increased production	Improved sterility	Advanced control
	Constant media charge	Batch scheduling
	Fewer upsets	Interactive control
	Reduced downtime	Predictive control
	Process optimization	GMP monitor/logging
	Improved rate or yield	
Improved product mix	Profit optimization	Optimize cycles
	Scheduling	Rapid changeover
	Inventory control	
Improved quality assurance	In-process control	Reduced errors
	Hazards control	Reduced human intervention
	Automated assays	GMP monitor/logging
	Data logging	Manufacturing specs
	Reduced giveaway	
Lower maintenance cost	PM scheduling	Equipment file
	Vibration analysis log	Work scheduling
	Predictive repair	Reduce spare parts
Improved safety	Interactive alarms	Automated sensors
	Impersonal monitor	
	Logging of problems	
Process knowledge	Creation of algorithms	
	Reduce human variability	
	Modeling	
	Expert systems	
Reduced energy consumption	Improve boiler efficiency	
	Optimize compressor use	
	Optimum mix for cooling	
	Temperature programming	
	Maximize recovery	
	Load shedding	

Barsamian discusses plant-wide computer control.[187] He presents some generalized cost information. In a large control system (600 control loops, 10,000 tagged items, 5 operator consoles), computer hardware and system software make up about 17% of the overall cost. If instruments and basic controls are already in place, cost breakdown is

	% of total cost
Application technology licensing	23
Computer hardware	10
Computer software	7
Vendor engineering	12
Control house revamp	10
Application engineering	12
Project management/engineering	10
Analyzers	7
Contingency	7
Training	2

It is quite possible that, if all above expenditures are required, $300,000 of computer hardware would result in a total expenditure of about $3 million. Even if some of the necessary parts of the project are done in-house, a cost is incurred. More importantly, a great deal more time might be required to achieve the goals established when the hardware was first purchased.

V. STERILITY

It is difficult to assess the cost impact of sterile design. If we assume that materials of construction would be unchanged in a "similar" plant, then sterile design cost increments would be found in:

1.   Need for air or liquid filters of special design
2.   Steam lines, piping, and valves to insure sterilization as well as special seals (add increment for added steam generation)
3.   Internal and external finishes to insure cleanliness
4.   Design of air compressors, including dehumidification
5.   Added insulation and/or tracing of lines
6.   Clean room design for inoculum and early seed development
7.   Isolation requirements for both fermentation and isolation (product/process dependent)

Sterile design has as its goal the elimination of all but the desired microorganism in a fermentor and the prevention of aberrant organism entry. Sterility is a probability concept; absolute sterility can be designed into the process, but it is seldom cost effective. Presence of a contaminant at a low dilution means absolute sterility has not been achieved; however, if there is no impact on biosynthesis, the empirical meaning of "sterility" has been achieved. Control of sterility assumes a major portion of fermentation design input and therefore impacts capital and operating cost. The rationale for sterility is not always clear. Most fermentation media are rich in substrates and cofactors for a host of microorganisms. Operating temperature, pH, pressure, and degree of aeration would satisfy very many microorganisms. Even if inhibitory substances are produced later in the cycle, there is an early growth period where desired microbial mass is very low. There exists, therefore, an incubator for development of numerous microorganisms. Only one organism is desired. Competitive organisms — even if they could not completely outgrow the desired culture — will utilize medium constituents (lower efficiency), change pH or redox, produce undesirable metabolites, and will generally lower product yield.

Many producing organisms, however derived, are no longer suited to growth or maintenance in nature. By selection or genetic transformation, the organism is either very fastidious or is "defective" when compared to wild types. Growth rate may be slightly lower than a related wild strain; small differences become very important in a competitive situation. Any contamination, however "slight", has the potential for product loss at best or batch loss at worst. Not every contaminant in the fermentor stage results in processing problems in extraction. However, there are few published cases where a contaminant has *aided* downstream processing. (One must exclude an enzyme, culture, or virus added specifically to perform some desired processing step; purposeful addition negates the concept of "contaminant".) Presence of a contaminant may have a detrimental effect on broth processing either due to presence of undesirable moieties or due to potential for destruction of desired product.

There is a series of principles which must be incorporated into design of sterile operating facilities. Among these are

1.   Discontinuities, sharp pipe turns, pockets, indentations and dead legs in line, vessel

or valve are undesirable. Solids deposition, air pockets, or stagnant layers can contribute to incomplete sterilization.

2. Positive pressure should be maintained within a sterile piece of equipment or a sterile space, as a clean room. Leakage (if any) outward is the preferred mode. Development of a partial vacuum might lead to leakage of nonsterile gas or liquid.

3. Microorganisms can migrate against a pressure gradient if there is some channel (as a fluid or porous solid layer) for movement. It is best to eliminate potential for any such occurrence, even if rate of migration is low.

4. Welded connections (smoothed internally and externally) are preferred. However, a combination of welded and flanged fittings will be used to allow easy disassembly for maintenance and repair. Threaded fittings that are not back welded are to be avoided in sterile service. Threaded fittings within the fermentor are to be avoided.

5. Nonsterile and sterile sections of a system should never have a permanent connection. Connections for cleaning of nonsterile water or medium transport should be temporary and should be disconnected prior to sterilization.

6. Any material that contacts a sterile stream (gas or liquid) should be nonporous, heat sterilizable, easily cleanable, preferably impervious, and long lived. Special attention should be paid to gaskets, diaphragms, liners, seal faces. Acceptable materials are Teflon®, high temperature elastomers, silicone rubber, glass, and certain ceramics.

7. Design should allow for double block protection. That is, separate sections of a plant should be sterilizable without interfering with ongoing operations. Sterilize by passing steam past one open valve in a line to a closed valve and steam trap; when heating completed, close first valve, open second, and sterilize by steam flow back to the first valve (also equipped with a trap).

8. Backflow into a fermentor of collapsed foam that may enter an air discharge line may present problems. Discharge points should be designed for ease of inspection, cleaning, and sterilization. Backflow preventers should be considered on liquid lines.

9. Steam should be introduced at high points and condensate takeoffs should be installed at low points. Any blanked section, depressed line or loop, should be piped for steam input or condensate discharge.

Other design criteria to improve sterility control are

1. Eliminate intermediate bearings for agitator shaft.

2. Be certain that any equipment (as separator) on air exit line can be cleaned and sterilized.

3. Air line can be introduced through side, near bottom, of fermentor. There is no need to run an air line into the top head and down through culture medium to a point below the bottom impeller.

4. Polish vessel internals and welds. Use closely coupled fittings and valves.

5. Permanent piping, jets, or rings within the fermentor for periodic cleaning should be avoided. Instead, a system that can be inserted easily, then removed, should be used.

6. Design coils for ease of cleaning and minimize deposition points. Avoid internal ladders and platforms if at all possible. Avoid internal lap joints which may trap liquids or solids.

7. Any upturn in a sterile line (especially air input section below impeller) should be provided with drainage holes or condensate piping.

8. Internals should have no pins or shims that might act as a solids collector.

9. Any instrument holders (as for electrodes) should be sloped to drain. Seals or O-rings should be located near the outside end of the support, thus, preventing an annular pocket within the support after electrode insertion.

10. Rising stem valves are poor choices for sterile applications. Any mechanical part which moves from a nonsterile environment to a sterile one presents a hazard. Diaphragm or plug cock valves are preferred choices. Valve operator does not contact the sterile stream. Even preferred valves can be drilled and tapped for steam and condensate lines.
11. Lines should be sloped slightly to drain. Valves should be installed to avoid any holdup. Vessels should drain completely. Any solids deposition is potentially hazardous to sterile operation.
12. Manifolding is acceptable, but must be reviewed carefully to insure that backup or cross-contamination does not occur.
13. Stuffing boxes are relatively inexpensive, but difficult to maintain. Spring-loaded seals or mechanical seals (high temperature at faces) are preferred.
14. In selected cases, it is necessary to pasteurize or sterilize fermentor contents or effluent air. Provision must be made to insure desired process result.
15. Use of a stator near one or more impellers is to be avoided. Such a device is difficult to clean.

There are a number of methods which can be used to batch sterilize liquid medium. (Some exotic procedures — use of sterilant gases or various forms of radiation — will not be discussed.) Indirect heating (coils, hollow baffles, jacket, external heat exchanger) can be used. Steam sparging (with attendant volume increase) is possible. The key criterion is that a sufficiently high temperature be employed for a sufficiently long time so that all organisms and spores are destroyed. With particulates present, resistances to heat transfer must be considered. Disadvantages of batch sterilization are inefficient use of fermentor (long heat-up and cool-down cycles), discoloration of medium is possible, loss of desirable heat-labile components, undesirable side reactions at long holding times. If sterile feed is required, the feed mixture and hold tank must be sterilizable as well as the fermentor.

Continuous sterilization provides remedies for many problems encountered using batch heating and cooling. In continuous sterilization, equipment external to the fermentor is used to heat, hold, and cool. Energy recovery via regenerative heat exchange is readily designed into the system as are control features. The process is less labor intensive than batch sterilization. Because much shorter residence times are possible, even at higher hold temperatures, media degradation is less likely. There is a potential for product yield improvement.

For clear media, especially where volumes are relatively low, sterile filtration is a reasonable choice. New and effective filter media have been developed that are sterilizable and cleanable and so can be reused. They can be tested to insure that removal characteristics are unaltered. Sterile filtration has been used in plant or animal cell culture to avoid nutritional loss in media which may result on heating. Special holders and associated equipment are supplied by manufacturers for plate or disk-type devices or cylindrical cartridge units.

Mathematical formulations for both batch and continuous sterilizers are available.[188] Design of sterilizers and other elements in a fermentation plant are covered by Kinoshita[27] as noted earlier. In a somewhat older reference, Parker presents a number of design criteria, gives simplified diagrams for practical solutions to potential contamination problems, and describes various fermentation procedures in a stepwise fashion.[189] Pym considers the "sterility requirement".[190] A probability of a contamination (say 10^{-3} infections per 8000 operating hours) is set and thermal death equations for various configurations are used to calculate sizes, retention time, and temperatures. Pym considers contaminant levels 10^3 to 10^9 organisms per milliliter for liquids and 10^3 to 10^4 organisms per cubic meter for air. Even with calculated values and laboratory results on thermal death rate, a certain amount of overdesign is included.

Sterile design should include a number of check lists. One such list should include equipment elements that are checked after every harvest, a second should include a weekly or monthly inspection, and a third should involve a thorough internal inspection as during down-time for preventative maintenance (frequency of once or twice per year per fermentor). Some companies include an internal inspection after *every* batch; others do not open the fermentor vessel except for special situations and perform exterior inspections on a routine basis. The details listed above are for routine inspection; a special inspection list (or even a PERT chart) should be prepared for a postcontamination vessel and auxiliaries check. Such a special routine should be dependent upon severity of contamination, contamination frequency in a given vessel, and even type of contaminant.

Preparation of a fermentor inspection check list in parallel with fermentor and instrument design is a good idea. One can conclude, in general, that an installation that can be easily inspected probably is well designed. Hidden areas are problem areas, again speaking in general terms. Design of smooth, uncluttered surfaces or probes that are exposed to water or cleaning solutions will also be amenable to visual inspection. In the course of a fermentor inspection, these items should be covered (and a positive record kept whether or not adjustment or maintenance needed):

Shell	Agitator seal
Shaft (keyways)	Drive (leaks, alignment)
Impellers	Valves (inputs and effluents)
Sparger	Exhaust point (air)
Coils	Sample line, valve
Ladders	Inoculum header
Foam probe	Defoamer valve
Exhaust line	Nozzles
Internal supports	Sight glasses
DO_2 probe	pH electrode

Leak testing is not performed on a routine basis. If contamination frequency is high, the test should be performed. Such a test should be performed during a preventive maintenance shutdown.

The importance of inspection and the emphasis on sterility can be shown by a review of typical turnaround time after completion of a batch. Values below do not refer to a specific process:

	Hr
Harvest (transfer to harvest tank or extraction)	2—4
Fermentor wash	2
Inspection	1—2
Dry sterilization	2
Charge medium for next batch	2

As noted earlier, the cost impact of sterile design is difficult to quantify. For simple conversion of carbon steel equipment to 304 stainless, the purchased cost factor is 1.82 for pumps and about 2.5 for other items. For carbon steel to 316 stainless, the cost factor is 2.0 for pumps and about 2.86 for other items. For heat exchangers (carbon steel equals 1.0), units with a carbon steel shell and 304 stainless tubes would cost 1.67 times more, and if 304 stainless shell and tubes were employed, the factor would again be about 2.86. These factors cannot be used for an entire fermentation plant, since (1) carbon steel is used wherever possible (nonsterile service, utilities, raw material storage, and so forth) and (2) certain isolation steps would require higher cost alloys no matter how product was synthesized. If a "conventional" fine chemical were to be made and isolated by organic/inorganic syntheses followed by separation, crystallization, and drying and the "conventional" plant were given a certain base cost, one could estimate that sterile design features of a fermentation plant would add 15 to 25% to delivered equipment cost. Of course, one assumes no changes in volume or capacity of key equipment items. Need for any supplemental utilities (air, steam) would require a separate and significant additional cost. The cost increment may be somewhat conjectural; however, the design features discussed do add a significant cost to design, purchase, and installation.

VI. TIMING AND PROJECT MANAGEMENT

At some point in the project cycle, one person should be selected to oversee successful implementation. The title may be project manager, project director, or any reasonable modification of these names. It is the responsibility that must be carefully defined, whatever the title. With responsibility comes the requisite authority. The size or importance of the project determines the level of the individual; further, the project director may or may not be the start-up manager. The choice is dependent upon size and scope. What is critical is that one person understands the corporate management objective and is given the responsibility and authority to execute it. Demonstrated competence in handling similar processes and excellent communication skills are two necessary, but not sufficient, attributes.

Project management is not a skill to be learned by the project director while on the job. Experience in project management should be learned in various subordinate roles of gradually increasing complexity. Project management is a rather specialized technical and personnel skill. If the project is small in scope and cost, the project director may assume many or all of the responsibilities of project management. For a grassroots plant, separate persons will be required for each position. Project management is covered in many publications. One reference details a number of tasks or responsibilities:[191]

- Product design
- Process planning
- Allocation of resources
- Facility location
- Production planning
- Materials management
- Scheduling
- Human resource planning
- Quality control
- Maintenance management

The project manager (as well as the project director) need not be the lead person in each subproject. In most cases, expertise in a specific area resides with others. Overall project execution requires the team leaders to understand what, why, and when tasks are underway and how and when they will be completed. Perhaps the greatest effort will be expended in allocation of personnel resources and establishing necessary priorities to facilitate project completion.

Project timing, with major time sequences shown, is described in Figure 20. Time is in arbitrary units as is extent of overlap. Length of each subgroup as well as overlap depends upon the size, complexity, and nature of the project. The start-up is a complex and involved affair (even if only a modification to an existing plant is made) and this important interval is covered in a separate section below.

175

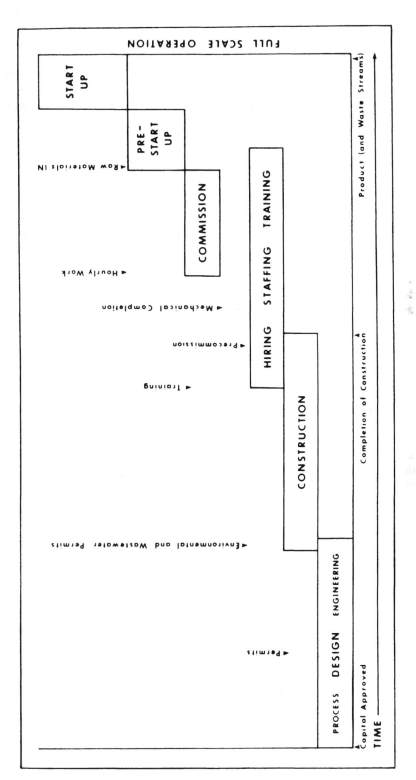

FIGURE 20. Project timing.

Table 37
PROJECT REVIEW AGENDA

Project schedule	
Scheduled labor	Engineering design
Overtime	Procurement
Added costs	Cash flow projections
Probable final	Category cost report
Percentage completion	
Raw materials	
Cost	Product specifications
Supply	Hazards (data sheets)
Quality control	
Start-up program	
Review PID	CPM
Vendor manuals	Check off
Testing	Training
Interrelationships	
Miscellaneous	
Scope changes	Flushing/cleaning
Utilities	Milestone schedule
Loading/warehouse	Project schedule (plot)
QC and testing	
In-process sampling	
Next meeting	
Assignments	
Forecast	
Schedule (vs. plan)	
Permits	
Procurement	
Construction (include all auxiliaries)	
Product specifications (especially changes)	
Market update	
Regulatory update	

Once the project leader is selected, he or she should assemble a team that can oversee the projected task. Project review meetings must not only be scheduled, but actually held. Careful minutes must be prepared in readable form; appropriate dissemination is essential. An outline of topics for such a meeting is given in Table 37. Emphasis may vary, but if items are eliminated routinely, it is likely that problems will ensue. The monthly report must have cost control as one of its critical concepts.

There is no ideal format for a monthly report (and attendees at the project review meeting may or may not be those that attend the monthly review). By their natures, the project review and monthly review are different. In general, the former is macroscopic in nature and is geared to major changes within or without the company that will have impact on the project. The monthly report is a combined update/checklist of the ongoing effort. One expects that minor alarms (in the form of delays, overruns, accidents, etc.) are monitored and appropriate planning accomplished to prevent these problems from growing into major disasters. The monthly report is especially useful for monitoring costs and hours vs. budget. As noted, cost control and timely execution are the major goals.

The monthly report will include a status report on:

Summary of project situation
Costs (vs. plan)
Cash flow

There will be more detailed emphasis on (and changes and exceptions in):

Engineering and design
  PFD and PID
  Civil engineering
  Architectural design
  Mechanical
  Piping
  Spare parts
  Modeling
  Electrical
  Instrumentation
  Computer (software and hardware)
  Safety/fire protection
  Dusts (hazards)
Forecasting (including use of contingency)
  Project cost (by category)
  Cash flow
  Cost trending
  Construction
  Engineering manpower
  Milestone schedule
  Project bar chart
Subcontracting
  Shop fabricated pipe
  Bulk piping
  Chutes and ducts
  HVAC
  Piping installation
  Mechanical installation
  Galleries
  Fire sprinklers
  Lighting
  Control room
  Insulation
  Electrical
  Painting
  Paving

Management of the project does not entail following a rigid set of guidelines. Successful project management involves judicious input to pursue optimization. Clearly, such input must recognize constraints of time and money. Still, it is possible to impact overall project cost and operation of the process in positive ways by appropriate review and suggestions. One such approach to optimizing engineering (meaning, in this case, contractor) input is given by Guidoboni.[192] While the discussion is centered upon ethanol production from biomass, there are lessons to be applied elsewhere. An analysis of published data shows that the elements that go toward determining ethanol selling price can be grouped and detailed (as percentages) as follows:

	%
Raw materials	72
Utilities	6
Labor	3
Overhead	2
Taxes/insurance	1
Depreciation	3
Sales/R&D/admin.	4
Taxes/profit	9
	100

Capital-related charges account for 6% of the total; by some rational extension, perhaps some portion of the labor and utilities charges may be capital related. In any case, the percentage is relatively small. Still, the four areas of the plant can be approached systematically in an attempt at significant engineering improvement. These are

1. Feedstock generation
2. Feedstock preparation
3. Fermentation
4. Product recovery

The feedstock areas that may allow substrate improvement or other economies are

Solids removal   Removal of inhibitors
Sterilization/pasteurization   Impurity use as animal feed or fuel
Cold sanitizing fluids   Conventional equipment
Heat exchanger design

The fermentation areas that deserve consideration are

Continuous operation   Instrumentation (minimize)
Ethanol yield improvement   Minimum agitation
Air sterilization   Ethanol recovery from off gas
Lined vessels   Design to available skills

Product recovery areas that may result in capital savings are

Use of conventional technology
Design energy efficiency with local fuel cost in mind
By-product recovery and use, preferably as is
Instrumentation to fit locale and process
Inexpensive materials (locally replaceable)

The basic success of local contractors in Brazil (where the project is to be built) is that historical data and development are used. The plants are "adequate and appropriate" if not thrifty, frugal, and spare. Nothing is custom designed. There are now about 400 fuel alcohol plants built over the years in Brazil; almost all have been built with local contractors, designers, and fabricators. There are a few lessons to be learned from this article. First, the earlier the involvement of the project team, the greater the chance for cost reduction. Second, technology should be appropriate and adequate for the process and locale. Third, there are potentially dozens of suitable alternatives, which can save money, that deserve consideration before design is frozen and even later, during construction.

A specific reference to fermentation plant project management is also available.[193] The paper discusses ICIs SCP plant and John Brown Engineers involvement with the project. The contract between the parties was fairly unique in that risk and benefit (financial) were shared. A team approach was taken with contractor process engineers working in an integrated team with ICI personnel at the latter's pilot plant. Both parties agreed on precontract procurement for certain long-term delivery items. Contract was signed January 1977. Design was complete June 1978. Plant was ready for commissioning July 1979. Success of contract execution phase was the result of:

- Appropriate handling of topical general management problems
- Rapid incorporation of unique and changing biological requirements into design
- Dealing effectively with nonbiological technical problems
- Development of design methods to minimize construction time
- Design development to assist commissioning
- Cost control of materials

Major areas of technical risk were identified and assessed. Alternate concepts were designed and, in some cases, procured. This may be considered a form of insurance for timely start-up. If a novel and very valuable product is to be made, such expenditures should be considered very carefully. Construction and commissioning of the ICI project are reviewed briefly. This was a massive project involving one of the world's largest continuous fermentation plants.

One potential outcome of a high technology project involving major expenditure is a successful start-up with greater than expected sales volume and profitability. Another scenario involves major cost overruns and delayed or even nonoperation of a new plant. This latter possibility, distasteful as it is to contemplate, deserves serious attention. The situation is reviewed and analyzed by Davis.[194] It is worthwhile considering a worst case result from another perspective. A disastrous project does not arrive at that state overnight. A recognition of warning signs (of early symptoms) may be of paramount importance in applying appropriate countermeasures. Should deterioration continue, one should attempt to avoid panic and apply some logic to avoid magnification of the loss. Davis correctly points to the unhappy fact that, as overruns occur, project managers tend to cut in areas where benefits are least obvious: managerial and design overhead, process control systems and software, quality assurance programs, and test procedures. Invariably, these cuts merely exacerbate the problem and magnify the overrun. The cause of most serious overruns is recognized as inadequate initial design. Process control programming is another example where complexity and compartmentalized expertise combine to present many unpleasant surprises. The author suggests a thorough design or redesign review. Recosting and review of market potential (volume and price, which may be better focused than at project inception) will, at the least, allow for somewhat more rational decision making. Abandonment is an alternative; money spent cannot be the stimulus for future expenditure.

Completion of a project and plant start-up are both very satisfying tasks. At least one other major task remains. While it may not be as pleasant as the others, it is extremely important for the ongoing operation of the plant and for future projects of the company. The task is, of course, the final project management report. This is a rather complex and often lengthy document. It should be both retrospective and prospective. Results, achievements, problems, and solutions have specific applicability and interest to personnel operating the new/modified plant as well as more general applicability to those who will be involved in future projects. A partial listing of headings and content of the "final" report is given

Table 38
CONTENTS OF PROJECT MANAGEMENT
FINAL REPORT

Major subjects	Subheadings
Project history	Capital (appropriation) request
	Scope of project
	Scope changes
	Milestones
	Completion report
	Investment credit
	Insurance valuation
	Profitability (if available)
Cost analysis summary	Estimated vs. actual
	Percent
	Equipment
	Project
	Field overhead
	Engineering
	Use of contingency
	Analysis and comments
Analysis of engineering	Estimated vs. actual
	MH and cost per drawing
	Historical curve
	Engineering rates
	Problems and analysis
Manpower usage	Unit and total costs
	Analysis
Cost data	Historical cost statistics
	Final cost (by element)
	Equipment cost curves
	Craft labor rates
	Efficiency by craft
	Analysis (vs. projections)
Analysis of purchasing	Field work
	Home office
	Subcontractors
	Analysis (vs. projections)
Start-up	Details
	Cost

in Table 38. The discriptors show that history *and* analysis are to be presented. Highpoints should be those areas that went well (and why) and those that went badly (also with why, but without allocation of blame).

A review of such reports is essential prior to initiation of a new project of any magnitude. Lessons of the past can be used to prevent repetitive difficulties or disasters. If the lessons are never reviewed or disregarded, it is likely that the new project team will be forced to relearn them. The attendant delays and cost overruns will occur inevitably. For very obvious reasons, the final management report should receive very careful attention in preparation, in dissemination, and should serve as a key reference document.

VII. START-UP

One part of the start-up process involves the translation gap between research, development, and manufacturing. Once a decision is made to build or to introduce a process modification or a totally new process, certain events are triggered which, at best, lead to friction and conflict. At worst, millions of dollars are irretrievably lost due to lack of communication, misunderstandings, and simple obtuseness. Why does this happen? At the translation point (or better, "interval"), these objective and subjective factors come into play:

- Time pressure to meet product schedules
- Economic pressure to meet budget guidelines
- Fear of, and resistance to, change
- Fear of technological obsolescence (usually "recipient")
- Fear of technical failure (usually "generator")
- Language problems — semantic differences
- Ongoing design changes
- Extensive overtime
- Need to maintain ongoing production
- Quality problems (disposal requirements)
- Demand for product (sales organization and/or customers)

Each of these problem areas, whether potential or real, must be faced *prior* to actual start-up or process introduction. Definitions must be made and written which will explain over-lapping areas of interest and which will detail not only each task, but will delineate what is meant by completion of each task. A number of rules and guidelines can be used to circumvent or alleviate the problems. While many are obvious, they continue to be disregarded with enormous human and monetary cost as a result.

The start-up of any production plant is always a difficult and complex task; this is true, only more so, for a fermentation facility. In fermentation operations, the margin for error is often very thin; in many instances, there is no margin for error. The start-up, as both culmination and beginning, deserves a distinct focus. There is probably no other portion of the project cycle where the truism, "Time is money", is so apt. The start-up may encompass initial approval, design, and construction, but we will consider "start-up" as that period in the construction interval required to commission the plant and satisfy project criteria. Therefore, construction is generally close to complete and equipment and controls are in place ready for air/water batching and testing.

The beginning of the start-up predates final phases of construction by weeks or months. The following tasks (I to VI) must be completed prior to plant entry:

I.   Organization of start-up
        Staffing/organization — an organization chart is absolutely essential
        Budgeting — expense and capital items
        Training and safety review
        Timing plans with control diagrams
II. Project activity lists
        Schedule for each phase
        Meetings/communications
        Job analyses

The cost of start-up can range from 10 to 25% of total installed plant cost. Percentage may be lower for a plant that is a replica or that has extensive operating history for the process. Companies usually have tabulated criteria for creating a factored start-up cost based upon (1) complexity, (2) type of equipment, (3) experience factors, (4) dependence on other units, and (5) miscellaneous considerations (regulatory aspects, timing, geography). Greater detail on cost consideration for start-up is given in Table 39. Holland et al. present an equation

Table 39
START-UP COST CONSIDERATIONS

To start-up completion
  Expensed
    Normal production costs
    Living allowances
    Transportation
    Training (include early hires)
    Temporary hires
    Preparation of manuals
    Start-up supplies
    Consultant fees
    Vendor charges (assistance)
    Overtime
    Scale models
    Operational errors
    Off spec. product discard
    Consumed spare parts
    Temporary installations
    Engineering for design changes
    Minor construction errors
    Added laboratory load
    Research assistance
    Credit: product to inventory by products sold

During start-up and afterward
  Capitalized
    Major additions
    Major repair/replacement
    Material upgrade

Note: Most start-up expenses may be capitalized.

with many variables and estimates for these variables under differing conditions that estimate start-up cost.[195]

If there is one key personnel assignment, it is that of the start-up director or manager. The individual must be both sufficiently senior and sufficiently experienced so that he or she can be the primary decision maker and can allocate company resources without waiting weeks for approval. Both knowledge and clout are essential. The start-up is a period of great stress so that stability must be matched by considered judgment under pressure. Good communication ability is essential. The start-up team should have members from research, engineering, and plant operations. It cannot be emphasized strongly enough that the earlier in the project this team is formed and interacts, the greater the chance for ultimate success. There should be an advisory or staff group available to the start-up team. Such individuals would provide needed input on health matters, environmental issues, transportation, accounting, and any other issue where expertise is needed for a short-term response. Another important attribute for either team members or "consultants" would be clearly demonstrated competence in a similar position at a similar prior task. The start-up period is not a good time for training of team leaders. Lastly, experience levels of supervision and hourly work force must be considered. A shift schedule for the start-up must be prepared.

The start-up team should be identified. Task descriptions for each individual should be written. Decision makers must be identified; they should be on the start-up team and not two or three telephone calls away. Engineering and manufacturing personnel should be aware of, if not directly involved, with R & D effort on the project or process. Investment and design decisions must be thrashed out well before start-up and all groups must sign off.

Decisions should be reached at the operating level at the site. When decisions are reached at the general managers' or senior vice-presidents' levels and filtered back, the chances for success drop precipitously. Aim for joint accountability. The task force for start-up is not a part-time job. Work cannot be planned or executed on an "as time permits" basis. R & D, engineering, manufacturing, and marketing should be exchanging visits on a routine basis as part of the ongoing business of the company. This need not be routine or frequent, but two to four times per year seems reasonable. If these groups meet for the first time in 3 years at a point in time 3 months before a new process introduction, one can predict confusion, delay, and overplan costs.

It should be clear by now that the person with direct responsibility for the start-up (who may or may not be the project director) must have some rather unique skills.[196] At least four such skills are crucial. The person should:

- Be comfortable with ambiguity
- Be willing to take, and be comfortable with, risks
- Have a high degree of personal responsibility
- Be able to communicate freely and openly

The entire start-up will succeed if a flexible system is established and if responsible personnel are permitted (and encouraged) to find ways to complete tasks, despite explicit directions to use a formal system that is one of a host of barriers to implementation. A start-up situation is inherently nonstable and fluid; rigid or inflexible responses will magnify confusion and result in delay.

Another aspect of planning is often overlooked. While technical and engineering support is carefully allocated and personnel are made available, insufficient attention is sometimes paid to the laboratory staff required. During a start-up, the normal sampling load may go up by a factor of 10. Regulatory matters may require heavy sampling of critical control points within the process (especially purity and microbiological quality) and there will be concurrent high frequency sampling of effluents (air, water, solids) to be sure no violations occur. These samples are already additive to the extra load on process monitoring. Inoculum development and sterility testing will be difficult to schedule. Since the inoculum laboratory is new and laboratory equipment may be new, entailing a ministart-up of their own, some redundancy in capability is suggested. An example is a wet chemical backup procedure for an HPLC assay; another is a nearby outside lab with appropriate capabilities. Should the plant laboratory be manned by the usual number of personnel during start-up, one can anticipate a heavy overtime load. This, in turn, leads to fatigue and sample backup. Pressures on "getting results" build as subsequent processing steps may depend upon analytical results on earlier, in-process streams. The entire start-up may be delayed or even stopped because of insufficient laboratory manpower. A failure of the inoculum lab to prepare and test seed flasks stops the project. Early start of the inoculum lab (through seed flask development) is a good idea. Environmental testing of the lab can occur long before the start-up itself.

It has been noted that large organizations with a strongly hierarchical structure exhibit greater intergroup conflict than smaller companies. One reason may be the lack of continuing communication at the operating levels. Dialogue is filtered upward in each line organization; vice presidents may or may not communicate, but the middle level and operations persons rarely have an ongoing dialogue. The smaller company has an obvious advantage. There are fewer layers of authority; matrix management is not a theoretical construct — it is essential for the viability of the business. One or two persons may make decisions that cover the spectrum of investment to market introduction.

Once there is a detailed activity schedule prepared, commissioning and prestart-up plans are made, staffing and training are defined, and raw material and energy requirements known, it is possible to prepare a "start-up cost estimate". This is an update of a percentage figure used in preliminary cost estimates. The start-up cost estimate should be continually updated and, after start-up, a final critique should include analyses of predicted and actual start-up cost. The success or failure of the start-up is not guaranteed by the planning tasks; however, gaps or lack of care at this stage will almost assuredly result in larger and more serious problems during the next phases:

III.   Precommissioning
      Equipment checks and inspections
      Testing and construction quality review
      Equipment files
      Begin punch list and correction
      Lubrication program
IV.   Commissioning
      Preparation for operation
      Cleaning, flushing, lubrication
      Simulated operation (as with water)
      Charging of columns
V.   Activity schedules
      Prestart-up timing and sequencing
      Emergency drills

Mechanical completion is that point in construction when commissioning may begin; the event occurs at $90 + \%$ construction completion.

The equipment file on a project should be started even before vendors are selected. The first critical items that enter the file are

      Plot plans and floor plans
      Process flow diagrams (PFD)
      Piping and instrumentation diagrams (PID)
      Instrument/electrical schematics
      Overhead/underground distribution
      Equipment drawings

Once vendors are selected, equipment specifications and operating manuals can be included. The process and utilities manuals should be included. All project documents (research and development reports, monthly reports, schedule, inspection documents) should be filed. The entire start-up file should be kept as a separate entity. A purchase order file is needed. Other manuals include:

      Process manual (calculation bases)
      Operating instructions (with record and log forms)
      Hazards analysis
      Job descriptions
      Analytical procedures/specifications
      Engineering (calibration results must be included)
      Computer systems
      Inventory methodology
      Regulatory monitoring
      GMPs (including inspections) and GLPs
      Safety and emergency procedures
      Coding and special testing results
      Spare parts and stock lists

At some point prior to precommissioning phase and continuing on through the start-up, a punch list is begun. This is simply a documented method of recognizing and correcting planning/execution oversights in construction. Many items will relate to improper or overlooked contractor work; other corrections will be needed to bring the installation to design specifications. Everyone on the start-up team can and must contribute both as a member of the team and as a supervisor of a part of the organization. Examples of punch list items in a fermentation plant (but not limited to such a plant) are

Lines not sloped to drain	Correct steam pressure regulators
Improperly installed steam traps	Power to instruments
Shut-off valves and by-passes	Relief device installation
Tracing and insulation	Adequate lighting
Welds not polished	Necessary cleaning lines
Excessive drive vibration	Removal of exchangers/filter elements
Oil or water in instrument air lines	Safety hazards (pinch points)
Correction of leaks	Removal of blanks

The list is by no means exhaustive, but presents typical problems encountered during commissioning. A problem that one must deal with concerns design modifications or pet projects that are camouflaged as a punch list item. One person (usually in engineering) should be given responsibility for evaluation of punch list items. Once an item is approved that person should institute a standard method for correction. The other items which involve design changes must be referred to the start-up team or director for review.

In the commissioning phase, overall plant activities begin and shift work commences. Certain of the activities are given below as examples of what actually occurs:

1. Electrical power distributed to all locations
2. Sanitary, process, cooling water systems activated
3. Utilities (air compressors, boilers, refrigeration) started
4. Agitators turned on and power draw determined
5. Instruments on and calibrated (test alarms)
6. Computer on and phased in
7. Solids and liquid handling systems run and calibrated
8. Wastewater system started up and sampling started
9. In-plant communication system tested
10. All safety systems (fire control, showers, emergency power) activated and tested

A major training effort occurs in this phase; most should be planned, but a good deal is unplanned and is necessary, nonetheless, for successful, long-term operation. Certain signs should be monitored. Examples are record keeping (batch sheets, control charts, flows, print-outs) — is it adequate and does it respond to reality?; shift change logs — are they clear and complete and do people respond?; maintenance response — is the work order system in use and are priorities being set and followed? Any problems in communication not resolved here will reappear, probably magnified, during start-up and later operation. Another often overlooked point is that supervisory personnel, operators, and technicians that are on board at the beginning of start-up are not always there at the end. In other words, plan for attrition and retraining. The final step of the start-up is just that:

VI.   Process start-up
         Procedures

Introduction of substrates
Introduction of reactants, solvents, etc.
Complete start-up punch list
Quality checks
Performance guarantees
Start-up cost determination and critique

The start-up is complete when operating criteria (which must include rate, yield, and quality) established in the project justification are satisfied. A specific level of operation for a specific time interval is a usual criterion. A block diagram, showing sequence of events, is not only helpful, but will permit facile control of the start-up.

It should be clear by now that the start-up is a critical and nonroutine procedure where guidelines cannot be rigid and unpredictability is the rule. Nevertheless, careful personnel selection and equally careful planning can limit discord and focus response when and where needed. Events and performance during the start-up can be analyzed and a rather good predictor can be established which will forecast operational efficiency (hence, cost of goods) for full production of the plant.

The greatest challenge remains the identification of suitable products and services. There is still inadequate recognition of the fact that the science, however outstanding, represents a relatively small element in the commercial success of biotechnology.

Dunnill, P., *Chem. and Ind.*,
July 2, 1984

It is inconceivable that today's successful software and computer peripheral producers — or tomorrow's biotechnology companies — started with a careful analysis of ROIs on their research projects.

Schmitt, R. W., *HBR*, 63, 124, 1985

Before continuing, a reminder that the ROI result projected under this — or any other — method will be only as valid as are the input assumptions on which the result is based.

Peters, R. A., ROI — Practical Theory
and Innovative Applications, American
Management Association,
New York, 1979, 21

Chapter 5

PROFITABILITY

Research projects are initiated and continued, designs are created, and investments made to generate some monetary measure of "success". Profitability is a measure of the investment rationale. There are many intangible or qualitative benefits to certain investments, but there must also be an improved monetary position with time. Otherwise, the reason for, or execution of, the investment was faulty. How is profitability measured and what are the advantages and disadvantages of these various techniques?

A typical investment decision involves research and development costs, other preconstruction costs, total fixed capital investment, working capital, start-up costs, ongoing maintenance/operating investment, total annual expenses with inflation adjustments, selling price/volume projections, tax impacts, product cycle decisions, make or buy decisions, regulatory changes, and investment salvage value. There are probably other, minor variables. It is clear that every investment involves a host of uncertainties and sensitivities to changes in underlying assumptions. It is rather important to know — in a quantitative way — how various parameter changes impact project profitability. An analysis of changes in assumptions and resulting changes in profitability is called a sensitivity analysis. A knowledge of which inputs have important effects will direct effort toward greater control and knowledge of these inputs.

Even before profitability becomes an overriding issue, attention to the more mundane aspects of planning is needed. The specific requirements of the appropriation request have been discussed. The following is a very abbreviated list of planning and program needs:

Business plan
    Product
    Time frame for development
    Need, market size, sales growth — risks
    Manufacturing process and costs

>   In-house (capital needed) or coprocess
>   Funding structure (as R & D partnership or stock)
>   Expenditure, return, cash flow (at least 5 years)
>   Clearance procedure — regulatory requirements
>   Environmental impact
>   Personnel (organization and structure)
>   Patentability — local or global coverage
>   License capability (or desirability)
>   Ownership of rights (partnership or joint program)
>   Marketing plan
>   Channels of distribution
>   Strategy
>   Market share goals
>   Growth targets
>   Pricing policy
>   Profit targets
>   Product mix
>   Competitive analysis
>   Geographic distribution
>   Postmarket monitoring and surveillance
>   Reporting of physiologic/environmental effects
>   Advertising

While analyses of most of these issues are beyond the scope of this book, these subjects are listed to emphasize the fact that profitability or ROI is not merely related to investment, cost of goods, and sales revenue. It is possible to relate ROI and payout time to these variables (and this will be done), but the generation of revenue is a function of many other inputs. Those inputs or programs are noted above. It is the rare product that is picked up by anxiously waiting customers at the plant gates. For even such a product, a time comes — all too soon — when mere product availability is insufficient to generate appreciable sales revenue.

There are not a large number of references that discuss fermentation economics generally (meaning costing, cost breakdown, process economics, sensitivity). Two such references are available for an overview of impact items.[197,198]

There are many associated services (and related expenses) that are involved in determination of operating cost and/or in determination of long-term profitability. Some items are included as separate line accounts and some items are lumped in an overhead figure, usually as a percentage of investment or some other known or estimated quantity. It is possible, however, to detail the associated manufacturing and other business services and assign a cost and use figure to each one. This is probably a better idea for a start-up or smaller business enterprise; in a large corporation, associated costs can be spread over very large volume or a number of products. Even if a small company utilizes consultants or outside services, it is important to track and control such costs. With few products, the burden on each one will be significant. In Table 40, the associated services that are needed to maintain and drive the business are listed. Costing method for each item is suggested. It should be obvious that any suitable or realistic costing procedure would serve. The important point is that some realistic figure be included individually or lumped together; whatever method employed means that the associated cost is *recognized* as a needed outlay in running the business. The best figure is the exact figure. That is usually difficult to get, so routine updates are needed for a new business. As a business history evolves, better estimates will be generated. Required work load (as man-hours) will be better defined and rates will be

Table 40
ASSOCIATED SERVICES AND COSTING METHOD

In Plant

Costing method

Purchasing	Plant OH or direct salary, benefits
Personnel	Plant OH or direct salary, benefits
Engineering (local)	Plant OH or departmental charge
Accounting	Plant OH or departmental charge
Distribution	Plant OH or departmental charge
Warehouse	Lease or equivalent charge
Transportation	Quotations and related costs

Central Location

Selling	Percent of sales or direct salary, benefits
Marketing services	
Packaging	Standard rate
Consumer information	Standard rate
Creative functions	Standard rate
Consultant	Contract or time and materials
Legal/regulatory	Standard rate
Tax and insurance	Standard rate
Research and development	Standard rate, direct, indirect expenses
Engineering	Standard rate and direct expenses
Credit/financial systems	Proportional to time spent or percent of total cost related to frequency of use
Management information	Standard rate
Customer services	Proportional to time spent
Public relations	Proportional to time spent
Environmental control/safety	Standard rate and direct expenses
Licensing (in/out)	Proportional to time spent

known. Use of in-house capability or outside organizations will depend upon needs and costs. The make or buy decision that relates to an intermediate or starting raw material can be extended to many, or even all, associated services.

Profit can be measured on a daily basis. While a month or a year is a more usual unit of measure, profit can be considered a short-term concept. Profitability is a long-term concept and investment decisions are more rationally reviewed over a number of years. An 11- or 12-year cycle is often used. Since profitability is a long-term concept, a single measure is not adequate.[199] Financial measures are essential, but there are many paths which will give, as an example, a similar discounted cash flow rate of return. Cash flow, however, may vary. Timing of investment may vary. Intangible changes may be more likely under a selected scenario. It is best to use a multiperspective approach to profitability analysis. Minimization of risk and reduction of uncertainty may support selection of a less than optimum project, if "less" refers to a single financial criterion. A balancing of alternative views is strongly preferred to a single, rigid analysis. If profitability could be predicted by a single mathematical technique, no investment decision could ever be incorrect.

Analysis of profitability begins with the earliest germination of an idea that might result in a product (see Table 41). While the economic model initiated at this time is prone to large error, the production design or marketing plan is probably prone to an error of the same magnitude. Lack of certainty is a poor excuse for not developing a profitability model. It should be (but sometimes is not) understood that two kinds of errors are possible at this point. A very large margin and short development time may prompt a rapid plunge into a

Table 41

**PRODUCT AND PROCESS DESIGN AND PROFITABILITY
MODELS**

Phase		Profitability
Early	New product concept	Economic models
	Design and pathway	Sales/market model
	Feasibility studies	
	Alternatives, form	
	Surveys and forecasting	
Intermediate		
Experimental	Production	
(what?)	(how?)	
Preliminary design	Process selection	Manufacturing model
Prototype	Process studies	Sensitivity analysis
Development phase	Design selection	Refine sales model
Performance tests	Optimization	ROI
Modify, redesign	Manufacturing specs	
Packaging, dosage	Equipment selection	
Pilot plant	Site selection	
Quality, specifications	Resource requirements	
	Quality assurance	
	Design changes	
Final		
	Final formulation	Adherence to model
	Clearance procedures	Cost and margin
	Installation	Correction of model
	Start-up	
	Operation	
	Product sale	

novel concept. Enthusiasm and money spent may be very high at first; as the profitability model becomes focused, both enthusiasm and funds dwindle. The error of the second kind may be considered the bureaucratic mistake of doing nothing, i.e., economic models invariably show no or low profitability. By never venturing, nothing is lost. Of course, the large gain is overlooked as well. With these potential errors in mind, one progresses to the intermediate stage of answering "what?" and "how?". Both process design and development proceed simultaneously with refinement of the profitability model. The problems at this stage involve personal "investment" (time, prestige, effort). It is often hard to admit defeat after massive expenditure. There is a tendency to keep trying for that one important breakthrough which will salvage the product. The one key point here is that continuation of the project with serious quantitative misgivings may mean there will be an even greater loss of funds (and prestige) some months or years down the road. Since large investment decisions must be made at some point in this stage, a careful production and sales model must be prepared with an estimate of financial return. It is this model which is circulated for approval and which becomes the guidebook for the final stage and for commercial success or failure.

The model should be updated at or near start-up and it is very important that poststart-up adherence to the model be monitored. It is extremely rare that the profitability, exactly as predicted, is achieved. Whether performance is better or worse than expected, there are many good reasons for an annual "conformance" report on the status of the process or product. After a company has completed a few projects, it may be possible to recognize a systematic bias in preparation of profitability models. Is cost of goods always under- or overstated? Are sales plans too conservative, too grandiose? Are competitive pressures

Table 42
PROFITABILITY FORECASTING IMPACT FACTORS

Fixed capital investment	Sales volume[a]
Site selection	Sales price[a]
Construction time[a]	Inflation rate
Start-up time[a]	Income tax rate
Start-up costs	Interest
Depreciation method	Regulation[a]
Obsolescence[a]	Legislation
Maintenance costs	Environmental impact[a]
Working capital	Competition[a]
Patent protection[a]	By-product credit
Raw material availability	Salvage value
Raw material price	Calculated risk
Plant replacement	Contractual agreements
SGA/RD costs[a]	Product effectiveness[a]
Monetary exchange rates	Side effects[a]

[a] Factors which may have a proportionately greater effect in biotechnology-oriented projects.

considered properly? Were regulatory affairs considered adequately? While corrective measures are taken on the production underway, improved planning for new projects is assured.

A table of factors has been prepared that lists inputs which have a greater or lesser impact on profitability both in forecasts and in the real world.[200] They are listed here, with some additions, since planners should take cognizance of these factors and give each an appropriate weighing (Table 42). The factors are universally applicable, but will be weighed differently depending upon the technology and product involved. The well-known rule that "anything that can go wrong does go wrong" applies to this list. Usually, the unweighed or neglected factor causes the greatest pain. For fermentation or biotechnology, there are factors that have a proportionately greater impact on profitability; these are shown as footnote a in the table. While an element of personal opinion is involved in such a selection, the dozen factors indicated should receive extra or more intensive evaluation.

There are a number of ways of determining profitability. Table 43 lists four common procedures with advantages and disadvantages of each. Details, including formulas and specific methodology, can be found in a number of useful references.[111,201-204]

Since standard formulas can be used to generate various profitability measures (see Appendix A for definitions), computer programs are commonly used. The computer has simplified the technique, but has also led to a proliferation of calculations. The general methodology of establishing a project ROI will be given without step-by-step programming details. Information must be entered to generate outputs or reports such as: cost data, earnings forecast, present value analysis/discounted rate of return, and breakeven sales. If a suitable program is not available in-house, many commercial software packages exist. Invariably, a commercial program must be adapted to individual usage. Input data should be acceptable as numeric or nonnumeric. A convenient time interval must be selected; individual year data or summed results should be available. Items to enter are

Capital investment (year 0)	Raw materials cost (multiple products)
Ongoing investment (out years)	Manufacturing labor cost
Production values (multiple products)	Maintenance labor cost
Start-up cost	Indirect labor cost
Dismantlement cost	Contract maintenance

Table 43
METHODS OF IDENTIFYING ACCEPTABLE INVESTMENTS

	Average rate of return (ARR)	Payback method (PB) (also, payout time)	Present value (PV) (also, net present value)	Internal rate of return (IRR) (also, DCFRR)
Pro	Simple to calculate A single measure Related to performance Easily understood Related to cost of funds Allows ranking Adjusted for investment	Simple to calculate Emphasizes investment safety Easily understood Favors short investments Considers cash flows Scale independent Allows ranking Universally used Effective when cash is short	Reliable valuation Incorporates desired ROI A discounted cash flow method Relates capital cost to investment Commonly used Considers all cash flows	Very reliable Measures true ROI Related to performance Can be compared to cost of funds Discounted cash flow Adjusted for investment scale Allows ranking Includes cash flows over life of project
Con	Simple average hides data Investment basis ill defined Ignores timing of returns	Simple value hides data Does not measure profits May be unreliable guide May favor underdesign Does not include return after payback Does not consider time value of money Short-range bias	Precision may be deceptive Large data requirements Does not relate to year-to-year performance Not readily understood Critical to select correct discount rate	Precision may be deceptive Large data requirements Estimate costs of capital Does not relate to year-to-year performance
	ARR is ratio of average profits after tax to average investment	PB is a measure of time required until original investment recovered	PV is the maximum that can be paid for an investment and still earn a pre-set rate of return; discounting is done (usually) at the minimal acceptable rate of return NPV = PV ± immediate cash flow	IRR is the ROI that results in a present value equal to the cost of the investment; the present value of all cash flows is zero at the discounted cash flow rate of return

Spare parts cost	Supervisory costs
Applicable sales taxes	Payroll, tax, benefits
Office services	Maintenance materials
Depreciation	Operating supplies
Insurance	Electrical cost
Taxes (property)	Energy costs
Packaging supplies	Rental costs
Shipping labor	Royalties
Shipping supplies	Demurrage
Receivables (cost)	SGA cost
Inventories (cost)	R & D cost
	Escalation factors (each item)

Obviously, some personal selectivity is involved; not all items need be entered. Costs can be entered as:

1. Year-by-year item
2. First year plus an escalation factor
3. First year plus escalation, then ratio each upcoming year based upon production
4. Ratio first-year cost to investment, then escalation factor
5. Specify unit cost, ratio of units to a production unit, an efficiency of production, and an escalation factor

Price can be entered in similar fashion, but is not normally ratioed to units of production. Many entries are entered as a percent of investment or a percent of labor and supervision cost. Escalation factors can and should be different for different line items. SGA/R & D are usually given as a percent of sales. Companies have selected "standard" percentages for these figures. Receivables and inventory are entered based upon company practice and/or turnaround. Changes in inventory or sales year-to-year will mean that an incremental change in these costs occurs each year. The change may be positive or negative. It is not a good practice to neglect inventories and receivables. In general, large amounts of money are involved.

Let us assume that an 11-year interval is used for profitability analysis. A typical printout would be tabular in format with years 1 through 4 or 5 shown, years 6 or 7 through 11 summed, and a grand total. Lines would show dollar figures (in thousands or millions):

Year	Fuel
Product (1,2, ... n)	Utilities
Start-up cost	Office services
Construction tax	Depreciation
Raw materials (1, 2, ... n)	Insurance
Payroll charges	Taxes
Supplies	Packaging labor and supplies
Expenses	Shipping labor and supplies

The final line would be total cost of goods sold with a useful adjunct of total unit cost for each product. Programs exist for any division of future flow of funds (by quarter and by year for different future time periods). The selected program should allow for simple change in key input variables so that sensitivity to important factors can be calculated. The complexity of inputs is determined by corporate or individual requirements.

The earnings forecast would have added inputs of provision for income taxes (percent). The program would generate a table with these lines printed as before:

Year	Profit from operations (PFO)
Sales price (1, 2, ... n)	Income taxes
Sales units (1, 2, ... n)	Investment tax credit
Net sales (1, 2, ... n)	Net earnings
Total net sales	Gross plant investment
Cost of goods (total)	Receivables
Gross profit	Inventories
SGA expense	Total initial investment
R & D expense	Annual investment

The final line would add net earnings and depreciation (from another program) and deduct investment to show net cash flow. Cumulative cash flow, for the 11-year life, could be shown as well.

The final print out would require only insertion of discount factor. Cash flow in would be shown for each year; a discount factor for each year and a present value for each year would also be calculated and shown. Cash flow out would be shown with each year's figure discounted. Present value of the outflow is shown. Cash flow in over cash flow out is the present value return. Average earnings over the investment give the ROI.

The citric acid project that has been discussed throughout this book can be used in one such example. The first year printout of the base case would look like (see also Table 29, Chapter 2):

Year 1 (figures in thousands)

Production	40,000 (units are pounds)
Start-up expense	2,280 Units (and below) are dollars
Raw materials	4,976
Payroll	
Manufacturers' labor	900
Maintenance	4,560
Superintendant, office, laboratory	225
PTB	300
Sub	5,985
Operation supplies/expenses	570
Utilities	6,040
Other	
Depreciation	5,182 11-year interval selected (straight line)
Insurance	570
Quality control	180
Waste treatment	600
Real estate tax	1,425
Plant OH	1,980
Contingency	342
By-product credit	(600)
Sub	9,679
Total cost of goods	29,530 (unit cost is $0.738/lb)

Various assumptions, as described earlier, can be made for escalation of each line account. The computer program can print quarterly, yearly, or 11-year summation values. Eleven-year summation shows:

Total (figures in thousands)

Production	440,000 lb
Start-up expense	2,280 Units (and below) are dollars
Subtotals	
Raw materials	67,108
Payroll	85,028
Supplies	7,300
Utilities	77,359
Depreciation	57,000
All other	55,437
Total cost of goods	351,512 (average unit cost is $.799/lb)

The earnings forecast follows. Year 1 and 11-year totals are shown. All production is sold and there are no capacity increases or productivity increases. Selling price escalation is 5% per year. This may be optimistic, but any value (including zero) can be selected. Utilities escalation is assumed at 3% per year with the same proviso. Most available programs will allow for costing and sales of multiple products from one plant. The start-up is a one-time cost. While no ongoing capital investment was assumed, programs allow for investment on a yearly basis. SGA and R & D costs are not included; sufficient ongoing corporate efforts and expenditure are assumed.

	Year 1	Total	
Sales price, $/M	830	—	
Sales, lb	40,000	440,000	(000)
Net sales	33,200	471,665	
COGS	29,530	351,512	
Gross profit	3,670	120,153	
Prov. for income taxes	1,835	60,077	(50%)
(no investment tax credit)			
Net earnings	1,835	60,077	
Receivables (12.5% sales)	4,150	6,760	
Inventory (8.33% COG)	2,461	3,070	
Gross plant	57,000	57,000	
Cash flow, cumulative	(56,594)	50,247	

The present value analysis shows ($000):

Year	Net earnings + depreciation	Discount 10%	Present value
1	7,017	0.952	6,680
2	8,591	0.861	7,397
3	9,050	0.779	7,050
4	9,536	0.705	6,723
5	10,049	0.638	6,412
6	10,592	0.577	6,112
7	11,166	0.522	5,828
8	11,772	0.473	5,568
9	12,412	0.428	5,312
10	13,088	0.387	5,065
11	13,803	0.350	4,831

Table 44
SENSITIVITY ANALYSIS — CITRIC ACID PLANT

11-Year totals are shown

	Base case	Investment to $45MM	Production up 10%	Raw material down 5%
Production (000 lb)	440,000	440,000	484,000	440,000
Cost of goods ($000)	351,512	317,400	351,512	348,154
Net sales ($000)	471,665	471,665	518,832	471,665
Net earnings ($000)	60,077	77,133	83,660	61,756
Cumulative cash flow ($000)	50,247	67,596	73,154	51,957
Cash flow in/out	1.03	1.32	1.23	1.05
ROI, %	8.4	13.3	11.6	8.6

	Investment + 10%	Production down 10%	Raw material up 5%
Cash flow in/out	0.93	0.83	1.01
ROI, %	6.7	5.1	8.2

This gives cash flow *in* which is $66,979,000. Cash flow *out* (investment and working capital) is handled the same way with 11-year discounting. The cash flow *out* is $65,006,000. Therefore, the present value return is 66,979/65,006 or 1.030. The ROI is taken as average earnings divided by the investment expressed as a percentage: 5,462/65,006 × 100 or 8.4%.

This is not an especially attractive return. It would be distinctly unattractive for a concern having a culture and process, but little or no previous experience in fermentation. For a diversified pharmaceutical company with large fermentor capacity, the return on this investment must be compared to many potential alternative investments. It would probably remain unattractive unless all other potentials were worse. For a *producer* of citric acid, other factors enter the picture. What is the competitive situation? Is there a desire to preempt any potential competitor? What will this capacity do for market share? Is an investment needed to protect an ongoing business? How old are the existing plants? It may be that the alternative of doing nothing will lead to certain other ramifications 1 or 5 years down the road. A more reasoned approach may involve the concept of "differential ROI". That is, two scenarios are compared. The do-nothing alternative may lead to very unpleasant consequences in future years. The obvious point being made is that an ROI calculation, in itself, gives only one piece of information. This piece of data may be very important, but it is not the only value needed to make an intelligent investment decision.

The immediate question that arises concerns economic response to altered inputs. In other words, a sensitivity analysis is a normal accompaniment to a profitability calculation. Some alternatives are shown in Table 44. The base case has been discussed above. The major impact item in the table concerns investment. A reduction to $45 million increases ROI to 13.3%. Other changes are assumed and results shown. A sales price reduction of 10% gives the same present value return as production down 10%. Raw material price changes have relatively minor impact.

From these results, one may hypothesize a "best case" scenario. Investment would be $45 million. Production would be up 10% from the base case. Raw material cost would be down 5%, manufacturing labor would be cut from $4.56 to $3.60 million (year 1), selling price for year 1 would be \$0.913/lb instead of base case value of \$0.83/lb. For such an optimistic study, cash flow in/out is 1.85 and ROI is 21.7%.

One final study utilizes the "best case" scenario with additional expenses. SG and A expense is inserted at 7% of sales, escalated 5% per year. R & D expense is put at 5% of sales, again escalated at 5% per year.

	Best case	Best case (plus SGA, R & D)
Cash flow in/out	1.85	1.48
ROI, %	21.7	15.9

The examples shown present some typical approaches and are given to exhibit methodology. The assumptions may or may not be valid and each result must be qualified by what information was used (or not used) to generate performance results. A useful reference which gives a step-by-step approach and sample tables and calculations for ROI is available.[204] Many programs already have present value tables incorporated as subroutines so that any discount rate can be used. A rather complete compilation (to 20 years, from 1 to 25% compound interest rate) is available for the compound interest factor, the discount factor, the annuity future worth factor, the annuity present worth factor, and the capitalized cost factor.[205] Discount factors (1 to 20%, 1 to 40 years) are also available.[206]

A review of discounted cash flow (DCF) techniques, with certain modifications and subtleties added, is given by Powell.[207] Each procedure is defined and the mathematical formulation is given. Methods of establishing the "hurdle rate" or cost of capital are described. The importance of the risk/return relationship is stressed and the "capital asset pricing model" (CAPM) is discussed. There are 24 references to original articles and analyses. Definitions, formulas, pitfalls, and various options in evaluating return on investment are discussed in a number of chapters in a book on corporate finance.[208]

Profitability analyses need not be complex. With proper assumptions, it is possible to "guesstimate" return on investment. Further, well-known processes can be studied. Such an approach was taken by Mateles.[209] The improved cell yield from methanol at the ICI plant for SCP in England was used to calculate a productivity improvement from an assumed (but reasonable) base. With certain other assumptions, plus ICI published information, a revenue enhancement was calculated. The pretax profit differential is estimated at about \$1.26 million/year (it should be noted that this is based upon the plant's rated capacity of 60,000 t of SCP per year). With an assumed R & D cost of from \$0.5 to \$2 million, the ROI (pretax) is from 63 to 157%. No other capital investment was assumed and this is reasonable for a development that enhances a cell yield by 5%. After tax ROI is attractive. The point made is that profitability calculations can be made for the more mundane fermentation processes before embarking on a development project (or even afterward). Furthermore, appreciable returns are possible for even commodity chemicals. Manufacturing improvements are less newsworthy than expression of a highly active pharmaceutical in a novel host. However, such improvements can offer significant economic impact.

An appropriate conclusion to this section might be a restatement of the three quotations given as introduction. Biotechnology investment is similar to any high-tech investment; both risk and potential reward are high. Ideally, a portfolio of products should be under development so that risk is shared. An ongoing business should be capable of withstanding a product or project failure (that refers to termination in-house, prior to major investment or market introduction). Perhaps a series of project terminations might occur; the impact should not be catastrophic. Ultimately, however, investors and shareholders demand some return for their monetary outlay. The measure of risk and the selection of major projects will determine the success of the enterprise. The success is measured by an increase in value which may, or may not, involve some timely distribution of return. Ultimately, some monetary reward results. Quantitative measure of return on investment is essential even though many nonquantitative inputs are needed. There may be many measures of success in personal life or in scientific endeavors. For an ongoing business, achievement of an acceptable return on investment is the primary measure of success.

Chapter 6

CONCLUSION

The revolution in biotechnology which started some two decades ago shows no signs of abating. Not only is the rate of discovery accelerating, but the time span between first isolation and identification and volume production is being shortened. Most rapid development has occurred in the U.S. and Japan with some noteworthy advances in Europe. The three major reasons are not hard to fathom.

Research and development infrastructure — An important manpower pool was developed in Japan after World War II. Fermentation technology, both fundamental and applied, was transferred to Japan especially in areas of antibiotic biosynthesis. Developments in molecular biology, especially in academic institutions along both coasts of the U.S., are well known. A cadre of trained professionals, along with costly laboratory instrumentation and information science capability, exists in any area of the world that is at the forefront of biotechnology.

Manufacturing capability — Biosynthesis, isolation, and purification facilities are found where fundamental science unfolds and where market demand (and money) exists. New processes exhibit the production needs of very small tonnage products. Scale is more typical of a laboratory or pilot plant (Figure 21). Still, the existence of commodity fermentation plants presupposes the existence of laboratories, microorganism depositories, pilot facilities, and other support structures to support and improve production plants. It is not a complex matter to use pilot facilities, or scale-down, fermentation operations. Even though dosages are very small (pharmacological activity seems to be growing higher and higher per unit mass of product), major complexities enter the purification process. For certain of the newer products, 99.0 or 99.9% purity may not be high enough. However, analytical and small-scale separation equipment has been common for many years in the "biotechnology-developed" countries. It is this family of sophisticated devices that has been scaled to production use. Included in manufacturing capability are an understanding of regulatory requirements, quality control and quality assurance needs, packaging, storage and shipping, and sophisticated toxicological and other testing functions.

Sales, marketing, support services — Product emphasis is now on high activity, specificity, quality, and understanding of potential side effects. Sophisticated techniques exist to understand and segment the market. Target audiences need not have an abnormality or disease; targeted diagnostic probes are available to determine pregnancy, tendencies to certain normal or disease states, maximum period of fertility, and other lifestyle situations. Consumer response can be gauged before a product is even devised. Follow-up monitoring and technical services must support the product, once in commercial circulation. A rapid response, information infrastructure must exist along with distribution channels.

It appears from the above that the strong will get stronger. The absence of one of the many key "environmental" conditions needed for success in biotechnology may act as an impossible barrier to any future growth in areas lacking associated support structures. The corporate situation in biotechnology continues to change rapidly. Nature of the science, the technology, and the business are responsible for rapid shifts in strategy. It is often difficult to say whether one or another product falls into the area of biotechnology. Many corporations, both large and small, attempt to capitalize on the cachet of the term. That expectations as opposed to reality in the business of biotechnology are widely divergent should have been predictable. Regulatory requirements should have been well known to both scientists and investors. The therapeutic area requires heavy involvement with regulatory bodies and expenditures are very high. A 5- to 7-year time frame for approval would be considered

FIGURE 21.   Pilot plant fermentors (30 ℓ seed and 300 ℓ production units). (Courtesy of B. Braun Instruments.)

average. The race for approval has already begun and a number of agents are in human testing and use. Human interleukin-2 is in clinical trials. Human growth hormone is in use. Tissue plasminogen activator and tumor necrosis factor are under test. The potential for one family of agents, the lymphokines, has been estimated at $1.7 billion by the end of this century. These agents include interleukin-2, B-cell growth factor, colony-stimulating factor, γ interferon, and macrophage activating factor. The spectrum of disease states that might be impacted covers tumors, AIDS, leukemia, suppressed immune response, and inflammatory diseases. The market, obviously, is there. Some of the developmental projects (especially for Genentech and Cetus) are listed in Table 45.

Conventional wisdom is that the host of biotechnology-based firms (see Appendix C for a partial list as of early 1986) will undergo some major transformations. Financial pressures will cause the smaller ones to merge (with one another or with large, established concerns) or to fail. Many firms have succeeded in the research and development phase and are now facing more complex and more costly stages involving manufacturing, clearance, testing, and distribution. Size, capability, and market goals all vary considerably. It will be interesting to monitor financial changes, growth, and mergers over the coming years. While the generation rate of new start-up companies can be expected to slow, one can predict that the merger rate may increase.

Table 45

COMMERCIAL DEVELOPMENTAL PROJECTS (AS OF MID-1985)

Genentech		Cetus	
Human insulin	In market		
Human growth hormone	Pending FDA approval		
α-Interferon	Pending FDA approval	β-Interferon	Phase II clinical
γ-Interferon	Phase II clinical	Interleukin-2	Phase II clinical
Plasminogen activator	Phase II clinical		
Tumor necrosis factor	Preclinicals	Tumor necrosis factor	Preclinicals
Factor VII	R & D	Colony-stimulating factor	R & D
Human serum albumin	R & D Animal studies	Breast cancer immunotoxin	Preclinicals
Herpes I and II vaccine	R & D	Monoclonal antibody antiinfectives	Preclinicals
Hepatitis B vaccine			
Bovine and porcine growth hormones	Animal trials	Porcine scours vaccine	In market
Foot-and-mouth disease vaccine	Animal trials	Lymphokines for shipping fever	Animal market
Other corporate research and testing			
Interleukin-1 blocker		Hoffmann-LaRoche, Merck	
Tumor necrosis factor		Biogen	
Interferons and lymphokines		Immunex, Hoffmann-LaRoche, Biogen, Schering	
Monoclonal antibodies		Centacor, Cytogen, Johnson & Johnson, Hybritech, Damon, Lilly	
Hepatitis B vaccine		Merck, Biogen	
Tooth decay vaccine		Merck, Johnson & Johnson	

Adapted from *Bio/Technology*, 3, p. 605, 1985 and *Business Week*, p. 92, July 22, 1985.

The problems that exist in finding and producing physiologically active compounds remain difficult, if tractable. Some of the major issues are

- Identification of needed (high demand) bioactive moieties
- Identification of genes, vectors, hosts for expression; included are bacteria, yeast, plant, and mammalian cell lines
- Needed capital and other investment for equipment, plant, support structure, regulatory and postmarket approval, legal protection, liability, marketing, and sales
- Reproducible, high yield production, consistent and economic purification
- Appropriate return on large investment outlay
- Competitive pressures and the need for multiple product manufacture and sale; flow of new product discovery and use
- High rate of important scientific discovery with potential for product and process obsolescence

It seemed that the dawn of a new scientific and business aggregation was visible in the 1970s. The developments in biotechnology were, in fact, revolutionary. In many ways, we are still at the dawn. Perhaps we have moved an hour or two into the new day. It does seem that the rapidly changing environment in biotechnology has altered the earlier assumptions. Only now are the legal and regulatory issues being recognized and defined. The magnitude of the task of converting laboratory discovery into widely used product and, concurrently, building a new pharmaceutical industry is being recognized as well. The issues to be faced in coming decades may be more complex than the research problems faced in the last few decades.

While human disease receives the major focus, other product areas have not been neglected. There are many commercial opportunities in biotechnology, aside from human therapeutics. Important work, often including field testing and commercial development, has been noted in these fields:

- Microbial insecticides
- Emulsifiers
- Polysaccharides
- Enhancement of ice nucleation
- Sweeteners
- Preservatives and bacteriocides
- Amino acids, peptides, and proteins

A partial listing of opportunities and commercial ventures in biotechnology is given in Table 46. Obviously, not every project or product is at the same stage of development. In many examples shown, commercial products are available. In others, laboratory testing continues. Still, the potential is vast. Furthermore, the impact areas are growing. New ideas lead to opening of new opportunities. Once again, regulatory issues which seemed simple a few years ago have not been resolved. The nature of the science has caused many (perhaps most) of the knowledgeable persons involved to project an accelerating rate of use and application. However, others are disturbed by the potentials of the science and predict dire environmental and human consequences. It is true that there has been no serious environmental upset related to genetic engineering. One can extrapolate from the past, but that itself is no guarantee of certainty.

This book has covered many of the essential steps involved in bringing a fermentation product from discovery to market. The central theme has been economics, though many other practical matters have been reviewed. As discoveries are made, as new biotechnology products and processes are discovered and developed and as investment of manpower and money will be required, the concepts and steps covered here will have to be addressed. It is hoped that the alleviation of human want and disease will be one of the important results of the revolution in biotechnology.

Table 46
OPPORTUNITIES AND COMMERCIAL VENTURES IN BIOTECHNOLOGY APPLICATIONS AND SCREENS

Pharmaceutical

- Antivirals
- Antiprotozoals
- Antifungals
- Antibacterials
  - New precursors
  - Enzyme inhibitors
  - Potentiators
  - Target site direction
- Antitumor
  - Inhibition of metamorphosis
  - Cell wall components
  - Nucleic acid synthesis
  - Culture differentiation
- Test systems (probes and diagnostics)
- Psychoactives
- Fertilization inhibition
- Immunoactives (suppressants)
- Anticholesteremic
- Edema reduction
- Enzyme inhibition
  - Protein synthesis
  - Permeability
- Adenosine deaminase
- Transferases
- Alkaline phosphatase
- Carboxypeptidases
- Cell wall synthesis
- Fatty acid synthesis
- Elastases
- Chitinases
- Proteases
- Saccharases
- Glycosidases
- Esterases
- Blood pressure reduction
- Therapeutic peptides
- Enzyme modification

Agriculture

- Plant pathology
  - Phytotoxins
  - Phytoalexins
  - Plant diseases
- Plant growth stimulation
- Plant protection
- Insecticides
  - Toxins
  - Infective agents
  - Resistant cell lines
- Animal feed supplements
  - Vitamins
  - Amino acids
  - Protein
- Growth promoters
  - Rumen inoculum
  - Antibiotics
  - Growth hormone
- Herbicide resistance
- Nitrogen fixation

Industrial

- Polysaccharides
- Hydrolysis
- Methylation
- Isomerizations
- Specialty chemicals

Food

- Single cell protein
- Amino acids
  - MSG
  - Lysine
  - Phenylalanine
  - Aspartic acid
- Vitamins
  - Riboflavin
  - Vitamins B_{12} and C
- Organisms
  - Yeast (alcohol)
  - Dairy fermentation (cheese)
- Enzymes
  - α-Amylase
  - Isomerases
  - Rennin
  - Cellulases
  - Pullulanase
  - Proteases
  - Glucoamylase
- Polysaccharides
  - Alginates
  - Xanthan, gellan
  - Pullulan
- Nonnutritive sweeteners
- Shelf-life extension
- Flavors and fragrances
- Analytical testing
- Traditional fermented foods
  - Soy sauce, miso, tempeh
- Organic acids
  - Citric
  - Lactic
  - Acetic
- Fats, oils, surfactants
- Peptides, proteins
- Bio-probes for pathogens

APPENDICES

APPENDIX A
Glossary of Terms
(Investment and Economics)

Battery Limits

This term refers to total plant capital cost. *Inside* battery limits refers to plant capital related to equipment for the production process (manufacturing area). Such costs include fermentors, centrifuges, filters, evaporators, tanks, dryers, heat exchangers, packaging equipment, etc. *Outside* battery limits refers to plant capital related to auxiliary operations indirectly related to the production process. Such costs include electrical systems, boilers, cooling water, process water, refrigeration, warehousing, raw material storage, buildings, site development, pollution and environmental control, and interrelated piping for these services.

Total installed cost (TIC) is the sum of costs relating to the completely constructed plant. Included are estimates, engineering, purchasing, construction, start-up.

Book Value

Book value is current investment value, calculated as original installed cost less depreciation. In accounting, it is the net asset value. Gross book value is the original cost of the asset.

Break-Even Analysis

Break-even analysis shows the mathematical relationship among price-cost-sales volume variables. A key determinant of such an analysis is that sales volume that is essential to avoid financial loss; this is the point where cost of goods (dollars) equals sales revenues (dollars). As such, breakeven analysis is useful in risk determination, in pricing, and in production planning. The break-even *point* is that percentage of capacity at which income equals all fixed and variable costs at that specific level of operation.

By-Product Credit

Whenever any by-product (anything other than main or desired product) that must be disposed of can be sold for any purpose, recovered costs are included under this heading. Many such by-products become animal feed supplements.

Cash Flow

Net passage of money into or out of a firm, a plant, or a project as a result of operations is cash flow. Conventionally, cash outflows are *negative*. Examples are investments, expenses, costs, increases in outlays, increase in accounts receivable, and increase in inventories. Cash inflows are *positive*. Examples are reductions in outlays, increased profit, and higher sales (no higher offsets). Cash flow at the end of the economic life of a project is terminal cash flow.

NOTE: Depreciation is not a cash flow. However, cash flow is often defined as after tax net income plus certain noncash charges against income (such as depletion and depreciation).

Cost of Production

The *net* cost of production is the total cost of production less by-product credits.

The *total* cost of production is the unit cost of product including all elements of expenditure to produce the product. Elements included are raw materials, packaging, utilities, labor, supervision, fringes, depreciation, and overheads. Return on investment is sometimes included as a cost. See also variable and fixed costs.

Cutoff Rate

The minimum rate of return that is acceptable to management is cutoff rate. This value is preset by management and is utilized to weigh and compare investment options.

Decision Tree Analysis

A technique using a diagrammatic picture of a chain of choices and examines effects of future alternatives, outcome, and decisions that result from one or another present decision is decision tree analysis.

Depreciation

By U.S. tax code, depreciation is that dollar sum which represents "... a reasonable allowance for the exhaustion, wear and tear of property used in the trade or business, including a reasonable allowance for obsolescence." Depreciation is a noncash expense whose net effect is to provide tax-free income. Depreciation provides a method of retaining funds which would go to taxes; the impact is the depreciation value multiplied by the marginal tax rate. Reported profits can be increased by moving from an accelerated depreciation program to one that involves a straight-line calculation method. Regulatory authorities permit straight-line depreciation use for reporting purposes while concurrently using accelerated depreciation for tax purposes.

Depreciation, therefore, may be looked upon as a tax allowance or a cost of operation; its intent is to continually provide funds to finance plant replacement, in segments or in total. Replacement is rarely "in kind"; the fund is used for any optimal investment in the business. The depreciation "charge" is an accounting transaction and no cash expenditure is involved.

Discount Rate

The annual interest rate that is applied in compound interest calculations in investment determinations is discount rate. Also, it is used to refer to the rate of return (or expected financial reward) in calculation of present value.

Equity

Equity is the investment amount (dollars) in the company as provided by stockholders. Usually, the investment covers common stock only; however, in some instances, preferred stock investment can be included. Equity includes common stock and retained earnings; equity and net worth are equivalent.

Owners' equity is the excess of a company's assets over liabilities. Paid-in capital in excess of the par value of stock is also included in equity evaluation.

Fixed Capital Investment

Various terms and abbreviations are used to denote the sum of the inside and outside battery limits investment. One is fixed capital investment, another is total fixed investment, still another is total investment. This latter term should not be used, as the total investment should include cost of raw material inventory and cost of work in process and finished product inventory (see those terms).

Fixed Cost (See Also Variable Cost)

Those costs incurred in a manufacturing operation that are independent of the volume of production are called fixed. Fixed, or indirect, costs are incurred whether or not there is any output of product. Fixed costs which are obvious are depreciation, taxes, interest on investment, hook-up charges for maintenance of services (as well as minimum charges for these services), insurance, and funds expended to protect the facility. If a facility is kept operable, fixed charges can include much or all of labor and maintenance as well as supervisory and clerical costs to maintain a stand-by condition.

Indirect Cost

As has been noted under "fixed cost", all costs that are not directly related to raw materials, energy, and labor can be grouped as fixed or indirect cost. Other terms are factory overhead or plant burden.

Indirect cost also refers to the fixed capital investment. Such cost encompasses construction overhead, engineering expenses, taxes and insurance, contractors, home office expenses and fees, resident engineering costs, and contingencies. These values are not insignificant as a percentage of equipment cost and must be monitored and controlled carefully.

Intangible Assets

This special class of assets covers items that have no *physical* existence. Examples are goodwill, copyrights, trademarks, patents, and distribution rights. As with all assets, intangible ones derive their value from potential for generating revenue. In some start-up situations, intangible asset valuation may be greater than that for tangible assets.

Amortization is the process of writing off cost of intangible assets. There is some flexibility in estimating useful life of intangible assets; this means that valuation and amortization must be analyzed very carefully. It is possible to overstate net income by amortizing over the maximum allowable period.

Internal Rate of Return (IRR)

The internal rate of return is that calculated rate of discount which makes net present value equal to zero. One would normally accept an investment project if the rate of return (discount rate) is less than the internal rate of return. In other words, the investment has a greater potential reward than a known, standard investment such as a bond. The IRR is the return on investment that results in a present value *equal* to the cost of the investment.

Inventory

Inventory refers to quantity of goods and materials on hand. Each item on site has a cost assigned to it; the cost may be a simple payment (as for a raw material) or it may be a complex calculation (as for an in-process stage or finished good). Major inventories are for raw materials, work in progress (WIP), and finished goods; however, most plants of any size have significant inventories of spare parts, maintenance materials, and water or boiler treatment chemicals. Further, off-site warehouses may store a significant amount of any of these items. All should be considered and included in working capital determination.

Marginal Cost

Cost that is the incremental cost of making one extra unit of production without any additional investment in plant or facilities is the marginal cost, often called "incremental cost" or "differential cost". Another way of looking at marginal cost is as that amount by which total outlays are increased by producing the last unit of output at any appreciable volume of production.

Opportunity Cost of Capital

The opportunity cost is also known as the hurdle rate or simply the cost of capital. It is that quantity of money represented by the expected return that is foregone by utilizing funds in a given investment rather than placing the same funds in comparable (or in safe) financial securities. In theory, a well-managed company or portfolio will have investments that return *more* than the opportunity cost of capital.

Overhead

Term incorporates those items, in addition to wages and salaries, which are attributable to operating and supervisory personnel (foremen, shift leaders, etc.); included are all "fringe benefits" such as medical and dental plans, retirement, disability, vacation pay, payroll taxes, and social security payments. The items above are sometimes called *direct* overhead. Added to these costs are those entailed in servicing the plant, but not directly attributable to each unit of product. Such costs include administration, security, janitorial and grounds services, security and fire protection, safety equipment and uniforms, on-site medical services, cafeteria, telephone, and so forth. This latter grouping is sometimes called *general plant* overhead. Overhead is the sum of direct and general plant overhead.

Payback Period

That time interval which elapses that will return an original investment less the salvage value is the payback period. Normally, returns after the payback period are disregarded and the time value of money is not considered. Usually, it is taken after income tax.

Present Value

The current value of a stream of annuity payments is the present value. It may also be considered the value of a future payment discounted at the discount rate.

Future value = (Present value) $(1 + i)^n$, where i is the discount rate and n is the number of years. (The value i is an interest rate.)

The present value for investment decisions is defined as the expected payoff at a time period multiplied by $(1/1 + r)$, where r is the rate of return for that same time period. The rate of return (r) is also referred to as the discount rate, hurdle rate, or opportunity cost of capital.

The net present value is the present value plus (or minus) any immediate cash flow. An investment is considered a negative cash flow in this determination; net present value of an investment is its present value less the cost of investment.

Profit

Profit is generally defined as the excess of monetary returns over expenditure in one or more transactions. A time frame is usually involved. In a more philosophical sense, profit is defined as compensation to an investor for assumption of risk in some commercial enterprise. In this manner, profit is distinguishable from wages, rent, or interest.

In a quantitative sense, "gross profit" is equal to total sales less manufacturing cost. A time interval (month or year) is involved and manufacturing cost includes fixed and variable costs as well as depreciation. "Net profit (pretax)" is defined as gross profit less general expense for the same interval.

General expense includes all costs involved in maintaining ongoing transactions; examples are sales and marketing expense and administrative costs.

Replacement Value

The dollar value of a new facility (current dollars) to take the place of an existing, older facility is replacement value. There should be no increase in capacity, otherwise value must be corrected for expansion.

Return on Assets (ROA)

$$\text{ROA} = \frac{\text{profits (net, annual)}}{\text{total assets}} \times 100$$

Return on Equity (ROE)

$$\text{ROE} = \frac{\text{profits (net, annual)}}{\text{stockholders equity}} \times 100$$

where stockholders equity = total assets minus total debt.

Return on Investment (ROI)

The return on investment (ROI) is a measure of profitability generated relative to the investment required to produce those profits.

Simply,

$$\text{ROI} = \frac{\text{profits (net, annual)}}{\text{investments}} \times 100$$

ROI can be calculated in many ways (return on assets, return on equity, return on invested capital). Investment should include depreciable, nondepreciable, and working capital. Time value of money is not included in a simple ROI calculation.

Sensitivity Analysis

A mathematical technique is used to determine effect on a dependent variable when changes occur in one or more independent variables (or with changes in parameters which relate to the independent variables). One important example of such a sensitivity analysis is the separate or combined impact on project ROI of a reduction in anticipated selling price and/ or a reduction in volume (or units) sold.

Standard Cost

A projected and programmed estimate of cost based upon selected raw material costs and usages, standard rate and yield of production, and all allocated overhead costs is standard cost. Standard costs are what costs *ought* to be and are used to monitor and measure efficiency of an operation. Standard and actual costs may be identical, but most often they are not; if standard and actual costs are identical, variance is said to be zero.

Start-Up Cost

The nonrecurring costs that are expended after plant site completion and before that time when the plant operates at a continuing, acceptable level of output at planned cost is start-up cost. Expenses cover advance hiring and training, raw materials which do not end as salable product, overtime, minor equipment alterations, or modifications, as well as other added costs related to initiation of production.

Sunk Costs

These costs are those which have already been incurred in development of a project. They can be considered historical costs which cannot be recovered; they are irreversible outflows of funds. Sunk costs are irrecoverable and, therefore, irrelevant in reaching a decision to accept or reject a project (or its continuation).

Variable Cost (See Also Fixed Cost)

Inventory found in a manufacturing operation consists of raw materials and supplies, products not completed or in finished form (work in process) and completed, salable products (finished goods). Cost of inventories is the result of measuring the cost of resources which flow through the manufacturing process.

Variable costs are those costs incurred which are totally dependent upon volume of production. Raw material cost, cost of package or bag, and cost of expendable operating supplies are examples of variable costs. Energy that can be related to each unit of production is a variable cost. (Energy expended whether or not there is any production can be considered a fixed cost. In an operating facility, variable energy cost is usually higher than fixed energy cost.)

Work in Process

This refers to materials (usually abbreviated as WIP) that have been partly processed and have not entered finished goods inventory. While quantity is important, a cost is associated with WIP materials. This cost is directly attributable to all material, applied labor, energy, and overhead which can be applied to its production (albeit incomplete) at any intermediate stage in the overall process. WIP costing must include yield reduction to that process step.

Working Capital

This cost represents that portion of the total investment required to cover cost of raw material (feedstock) inventory, WIP inventory, and finished goods or product inventory. In some costing procedures, accounts receivable plus cash-on-hand less accounts payable is also included. A percentage of inventory and another percentage for WIP and FG inventory are used in determining working capital. Other components of working capital are supplies for product manufacture, prepaid expenses, taxes payable, and necessary cash to maintain business. Start-up expense is normally excluded.

Other costs which may be included are

1. Cost of handling materials to and from plant stores
2. Cost of inventory control
3. Readily available cash for emergencies

In the chemical process industry, working capital falls in the region of 10 to 20% of the value of the fixed capital investment. For a large pharmaceutical company, the range may be 15 to 30%. For a smaller, specialty biotechnology company, the ratio may be much higher, depending upon complexity of raw materials, intermediates, and value of finished goods.

Yield

In financial terms, yield is the overall compounded ROI on an investment. Yield can also be considered the internal rate of return of an investment.

APPENDIX B
Typical Fermentor PID

A typical fermentor PID is shown (Figure 22). Certain details are not shown; sizing of lines and valves is usually given as are materials of construction and some instrumentation detail. A key to symbols follows:

SLC	Low pressure steam supply	PI	Pressure indicator
AS	Air supply	TIC	Temperature indicator controller
PIC	Pressure indicator controller	TR	Temperature recorder
FR	Flow recorder	RV	Relief valve
FIC	Flow indicator controller	V	Vent

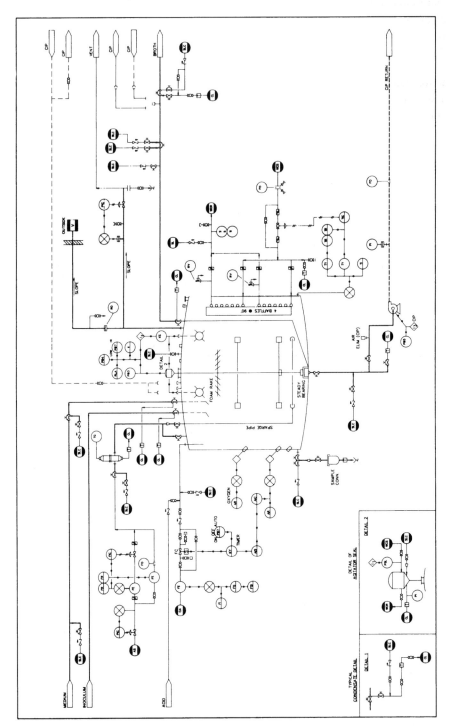

FIGURE 22.   Typical fermentor PID.

CL	Condensate	RD	Relief device
CA	Caustic supply	PSL	Pressure switch, low
WCR	Cold water return	T	Trap
WCS	Cold water supply	AR	Analyzer, recording
FO	Flow orifice	AIS	Analyzer, indicating
FC	Flow controller	AIC	Analyzer, controlling
FQ	Integral flow measurement	TY	Temperature relay
FE	Flow element	TG	Temperature gauge
XS	Limit switch	PB	Push button
RLA	Running light		

APPENDIX C
Biotechnology Start-up Companies

A list of biotechnology start-up companies is given. Not all companies are included; selection is somewhat arbitrary. Large, well-established pharmaceutical companies are not shown, except for acquisitions. Certain firms on the list are large and may be well established as well, but are generally considered part of the "genetic engineering start-up". It is expected that changes, mergers, and other events will date this list rapidly. Still, the extent shows that both financial incentive and technological interest are very high.

BIOTECHNOLOGY START-UP COMPANIES (INCLUDES MONOCLONAL ANTIBODIES, PLANT GENETICS, ANTIBODIES, DIAGNOSTICS, PHARMACEUTICALS)

A.B.I. Biotechnology
Advanced Genetic Sciences
Agdia
Agricultural Genetics
Agrigenetics (Lubrizol)
Agri Tech Systems
Allelix
American Bio Nuclear
American Biogenetics
Amgen
Applied Biosystems
Applied Genetics
Appligene

Bethesda Research
Biogen
Bio Chem Technology
Bio Logicals
Bio Polymers
Bioprocessing Ltd.
Bio-Response
Bio-Vec Technology
Biotech Research Labs
Biotechnica International
Bio Technics
Biotechnology General
Biotherapeutics
Calgene
California Biotechnology

Cambridge Bioscience
Cambridge Research
Celltech
Cellular Products
Centocor
Cetus
Chiron
Cistron Biotechnology
Codon
Collaborative Research
Collagen Corp
Cooper Biomedical
Cyanotech
Cytogen
DNA Plant Technology
DNA-X (Schering)
Damon Biotech
Diagnostic
Ecogen
Endotronics
Engenics
Enzo Biochem
Enzon

Enzyme Biosystems
Gen-Probe
Genen-Cor
Genentech
Genetic Diagnostics

Genetic Engineering
Genetic Laboratories
Genetic Systems (Bristol-Myers)
Genetics Institute
Genex
Genzyme
Hana Biologics
Hybridoma Sciences
Hybritech (Lilly)
Hygeia Sciences
IGI Biotechnology
Immunex
Immuno Modulators
Immuno Nuclear
Immunogenetics
Immunomedics
Immunotech
Imre
Ingene
Integrated Genetics
Interferon Sciences
International Genetic Engineering
International Plant Research
Invitron
Karyon Technology
Lee Bio Molecular
Life Technologies
Liposome
Microgenetics

Molecular Biotechnology	Plant Genetics	Syntro
Molecular Genetics	Porton International	T-Cell Sciences
Monoclonal Antibodies	Repligen	Tech America Group
Mycogen	Ribi Immuno Chem Research	Transgene
Neogen	SDS Biotech	Unigene
Neurotech	Seragen	Viragen
Oncogene Sciences	Serona	Wellcome Biotechnology
Petroferm	Summa Medical	Xoma
Phyto Dynamics	Sungene Technologies	Zymo Genetics
Phytogen	Synergen	Zymos

APPENDIX D
Periodicals and Reviews

There has been a proliferation in biotechnology publishing that has paralleled the business growth. A list of such publications is given. Once again, not all world or U.S. publications are shown and one can expect both additions and deletions with the passage of time. It is obvious that keeping abreast of technology is a difficult and time-consuming task.

PERIODICALS AND REVIEWS (APPLICATIONS, COMMERCIALIZATION, SCALE-UP)

Advances in Applied Microbiology
Advances in Biochemical Engineering/ Biotechnology
Advances in Biotechnological Processes
Agricultural Genetics Report
American Biotechnology Laboratory
Annual Reports on Fermentation Processes
Applied Genetics News
Biofutur (French and English)
Biomass
Bio Technology
Biotechnology Abstracts (Derwent)
Biotechnology and Bioengineering
Biotechnology and Genetic Engineering Reviews
Biotechnology in Japan Newsservice
Biotechnology Letters
Biotechnology News
Biotechnology Progress (A.I.Ch.E.)
Biotechnology Research Abstracts
Critical Reviews in Biotechnology (CRC Press)

Current Biotechnology Abstracts
DJM Enzyme Report
Developments in Industrial Microbiology
Enzyme and Microbial Technology
FEMS Microbiology Reviews
Genetic Engineering and Biotechnology Monitor (UNIDO)
Genetic Engineering News
Genetic Technology News
Journal of Applied Biochemistry
Journal of Applied Microbiology and Biotechnology
Journal of Biotechnology
Journal of Fermentation Technology
Journal of Industrial Microbiology
Membrane Separations in Biotechnology
Process Biochemistry
Progress in Industrial Microbiology
Topics in Enzyme and Fermentation Biotechnology (England)
Trends in Biotechnology (Elsevier)

APPENDIX E
Biotechnology Equipment Manufacturers

Since production and scale-up is an important subject, it is useful to have some indication of contacts for cost information, availability of equipment and reagents, as well as test and laboratory facilities to supplement in-house capabilities.

The listing which follows is selective and compiled from publications and advertising. Subheadings are

- Manufacturers (production-scale equipment)
- Reagents, adsorbents, peptides, antibodies
- Applications (laboratory, pilot plant, testing, instrumentation)

There is some overlap and many organizations can perform multiple functions. It is sometimes difficult to separate production-scale from pilot-scale equipment. There are engineering design firms and equipment manufacturers that are involved in biotechnology; however, these larger firms (not listed) are generalists and are less centered upon this area of technology. Many have biotechnology divisions. Their expertise and involvement are of a high order; their names and reputations are well known.

Manufacturers (production-scale equipment)

APV International
Abcor (Koch)
Alfa-Laval
Amicon
B. Braun Melsungen AG
Centrico
Chemap (part of Alfa-Laval)
DDS R. O. Division
Dominick Hunter
Dorr-Oliver
Imperial Biotechnology Ltd. (scale-up and custom production)
Koch Membrane Systems
LSL Biolafitte, Inc.
Ladish Co.
Millipore Corporation
New Brunswick Scientific Co.
Niro Atomizer
PCI R. O. Division
Pall Corporation
Pasilac
Pharmacia
Romicon
Separations Technology
Setric Genie Industriel
Westfalia Separator
Whatman

Reagents, adsorbents, peptides, antibodies

Applied Protein Technologies
Behring Diagnostics
Biomolecular Research
Bio-Rad
Biosearch
Bio-Yeda
Boehringer Mannheim
Boots Co. PLC

BSI/Serotec
Calbiotech
Charles River Biotechnical
Cooper-Biomedical, Inc.
Cruachem
Fisher Scientific
Gen-Probe
Hy Clone Laboratories
International Bio-Technologies, Ltd.
Janssen Biochimica
Life Technologies
OSC Laboratories
Oxoid, Ltd.
Pharmacia P-L Biochemicals
Pierce Chemical Co.
Promega Biotec
Rohm Pharma
Sheffield Products
Sigma
Tago
Vega Biotechnologies
Waters

Applications

(laboratory, pilot plant, testing, instrumentation)
American Bio Nuclear
Amersham
Amicon
AN-CON Genetics
Applied Biosystems
Applikon Instruments
Astra Scientific
BTC Diagnostics
Bachem
Beckman
Bellco
Bethesda Research Laboratories
Bio-Rad Laboratories
Biosearch
Bio-Tek
Bio-Yeda
Bioengineering Associates
Biotechnology Development
B. Braun
Brinkmann Instruments
Cerex
Charles River Biotechnical Services
Cole-Parmer
Cruachem
DNA Star
Endotronics

Forma Scientific
Gallenkamp
Genetic Design
HP Genenchem
Ingold
Ionics, Inc.
Isolab
Israel Institute for Biological Research
Karyon Technology
LH Fermentation
LKB Instruments
Lab-Line Instruments
Labsystems
Marubishi
Orion Research
Pandex Laboratories
Queue
Quidel
Radiometer
Setric Genie Industriel
Smith Kline Beckman
Systec
Technicon
The Enzyme Center
United States Biochemical
VG Instruments
Varex
Ventrex
Vir-Tis Company
Waters Chromatography
Wheaton Instruments
Yellow Springs Instrument

REFERENCES

1. **Anon.**, Industrial Research Institute's annual R & D trends survey, *Res. Manage.*, 28(2), 10, 1985.
2. **Anon.**, *Chem. Eng. News* p. 16, March 4, 1985.
3. **Anon.**, *Bus. Week*, special issue, p. 169, March 22, 1985.
4. **Anon.**, Cash flow: the top 200, *Dun's Bus. Month*, p. 44, July 1985.
5. **Anon.**, Chief scientist Schneiderman: Monsanto's love affair with R & D, *Chem. Eng. News*, p. 6, December 24, 1984.
6. **Ritchie, G.**, From discovery to commercial reality, some aspects of fermentation product development, *Chem. Ind.*, p. 403, June 17, 1985.
7. **Thomas, N.**, Decision analysis and strategic management of research and development: a comparison between applications in electronics and ethical pharmaceuticals, *R & D Manage. (U.K.)*, 15(1), 3, 1985.
8. **Schmitt, R. W.**, Successful corporate R & D, *Harvard Bus. Rev.*, 63, 124, 1985.
9. **Foster, R. N., Linden, L. H., Whiteley, R. L., and Kantrow, A. M.**, Improving the return on R & D, *Res. Manage.*, 28(1), 12, 1985; 28(2), 13, 1985.
10. **Szakonyi, R.**, Keeping R & D projects on track, *Res. Manage.*, 28(1), 29, 1985.
11. **Lilly, M. D.**, Problems in process scale-up, in *Bioactive Microbial Products 2*, Nisbet, L. J. and Winstanley, D. J., Eds., Academic Press, New York, 1983, 79.
12. **Bjurstrom, E.**, Biotechnology: fermentation and downstream processing, *Chem. Eng.*, p. 126, February 18, 1985.
13. **Bailey, J. E. and Ollis, D. F.**, *Biochemical Engineering Fundamentals*, McGraw-Hill, New York, 1977.
14. **Blakebrough, N., Ed.**, *Biochemical and Biological Engineering Science*, Vols. 1 and 2, Academic Press, London, 1968.
15. **Perlman, D.**, Fermentation, in *Encyclopedia of Chemical Technology*, Vol. 9, 3rd ed., Kirk-Othmer and Grayson, M., Eds., John Wiley & Sons, New York, 1980, 861.
16. **Blanch, H. W.**, Aeration, in *Annual Reports on Fermentation Processes*, Vol. 3, Perlman, D. and Tsao, G. T., Eds., Academic Press, New York, 1979, 47.
17. **Yoshida, F.**, Aeration and mixing in fermentation, in *Annual Reports on Fermentation Processes*, Vol. 5, Perlman, D., Ed., Academic Press, New York, 1982, 1.
18. **Humphrey, A. E.**, Biochemical engineering, in *Encyclopedia of Chemical Processing and Design*, Vol. 4, McKetta, J. J., Ed., Marcel Dekker, New York, 1977, 359.
19. **Sobotka, M., Prokop, A., Dunn, I., and Einsele, A.**, Review of methods for the measurement of oxygen transfer in microbial systems, in *Annual Reports on Fermentation Processes*, Vol. 5, Perlman, D., Ed., Academic Press, New York, 1982, 127.
20. **Votruba, J. and Guthke, R.**, On-line estimation of a fermenter aeration capacity ($k_L a$) by a computer and dissolved oxygen probe, *Biotechnol. Lett.*, 7, 267, 1985.
21. **Moes, J., Griot, M., Keller, J., Henizle, E., Dunn, I., and Bourne, J.**, A microbial culture with oxygen-sensitive product distribution as a potential tool for characterizing bioreactor oxygen transport, *Biotechnol. Bioeng.*, 27, 482, 1985.
22. **Aiba, S., Humphrey, A. E., and Millis, N. F.**, *Biochemical Engineering*, Academic Press, New York, 1986, chap. 8.
23. **Nisbet, L. J. and Winstanley, D. J., Eds.**, *Bioactive Microbial Products 2*, Academic Press, New York, 1983, chap. 1 to 5.
24. **Ritchie, G.**, From discovery to commercial reality, some aspects of fermentation product development, *Chem. Ind.*, p. 403, June 17, 1985.
25. **Dalby, D. K.**, Pure culture methods, in *Prescott and Dunn's Industrial Microbiology*, 4th ed., Reed, G., Ed., AVI Publishing, Westport, Conn., 1982, 44.
26. **Solomons, G. L.**, *Materials and Methods in Fermentation*, Academic Press, New York, 1969.
27. **Kinoshita, K.**, Fermentation equipment and instrumentation, in *The Microbial Production of Amino Acids*, Yamada, K., Kinoshita, S., Tsunoda, T., and Aida, K., Eds., Halsted Press, New York, 1972, 139.
28. **Kuenzi, M. T. and Auden, J. A. L.**, Design and control of fermentation processes, in *Bioactive Microbial Products 2*, Nisbet, L. J. and Winstanley, D. J., Eds., Academic Press, New York, 1983, 91.
29. **Lowe, C. R.**, New developments in downstream processing,, *J. Biotechnol.*, 1, 3, 1984.
30. **Van Brunt, J.**, Scale-up: the next hurdle, *Biotechnology*, 3, 419, 1985.
31. **Dimmling, W.**, Substrates in Fermentation Technology — Situation and Economy, paper presented in the 12th Int. Congr. of Microbiology, September 3 to 8, 1978.
32. **Florent, J. and Ninet, L.**, Vitamin B_{12}; **Perlman, D.**, Microbial process for riboflavin production, in *Microbial Technology, Microbial Processes*, Vol. 1, 2nd ed., Peppler, H. J. and Perlman, D., Eds., Academic Press, New York, 1979, chap. 15 and 16.
33. **Atkinson, B. and Mavituna, F.**, *Biochemical Engineering and Biotechnology Handbook*, Nature Press, New York, 1983, 1109.

34. **Perlman, D.,** Microbial production of antibiotics, in *Microbial Technology, Microbial Processes,* Vol. 1, 2nd ed., Peppler, H. J. and Perlman, D., Eds., Academic Press, New York, 1979, chap. 8.

35. **Brown, O.,** Sucrose and molasses as feedstocks for fermentation processes, *Chem. Ind.,* p. 95, February 7, 1983.

36. **Emery, W. L., Ed.,** *Commodity Year Book — 1984,* Commodity Research Bureau, Jersey City, N.J., 1984, 231.

37. **Hebeda, R. E.,** Syrups, in *Kirk-Othmer Encyclopedia of Chemical Technology,* Vol. 22, 3rd ed., Grayson, M., Exec. Ed., John Wiley & Sons, New York, 1983, 499.

38. **Anon.,** Cereals. I. Corn, *Feed Manage.,* 36, 12, 1985.

39. **Lewin, D.,** Political and economic aspects of soya bean meal markets, *Chem. Ind.,* p. 102, February 7, 1983.

40. **Williams, W.,** Oilseed production for oil and protein processing, *Chem. Ind.,* p. 603, August 1, 1983.

41. **Considine, D. M. and Considine, G. D., Eds.,** *Foods and Food Production Encyclopedia,* Van Nostrand Reinhold, New York, 1982, 623.

42. **Stringer, D.,** Economics of Strategic Raw Materials, presented at Biotech Europe 85, Geneva, May 21 to 23, 1985; cited in *Manuf. Chem. (England),* 56, 24, 1985.

43. **Dimmling, W. and Nesemann, G.,** Critical assessment of feedstocks for biotechnology, *Crit. Rev. Biotechnol.,* 2, 233, 1985.

44. **Milkie, R. C. and Clark, J. P.,** Use of Simulation Model in Design of a Multi-Product Plant, paper presented at A.I.Ch.E. Annual Meeting, San Francisco, November 29, 1984.

45. **Sha, P. Y. and Bohinsky, J. A.,** Structural design procedure for plant retrofit, *Chem. Eng. Prog.,* 18, 20, 1985.

46. **Soderberg, A. C.,** Fermentation design, in *Fermentation and Biochemical Engineering Handbook,* Vogel, H. C., Ed., Noyes Publications, Park Ridge, N.J., 1983, chap. 3.

47. **Charles, M.,** Fermentation scale-up: problems and possibilities, *Trends Biotechnol.,* 3, 134, 1985.

48. **Naveh, D.,** Scale-up of fermentation for recombinant DNA products, *Food Technol.,* 39, 102, 1985.

49. **Swartz, J. R.,** The large scale culture of microorganisms, *Am. Biotechnol. Lab.,* 3, 37, 1985.

50. **Gutcho, S. J.,** *Chemicals by Fermentation,* Noyes Data Corp., Park Ridge, N.J., 1973.

51. **Black, J. H.,** Risk and uncertainty analysis, in *Cost Engineering Planning Techniques for Management,* Marcel Dekker, New York, 1984, chap. 2.

52. **Clark, F. D. and Lorenzoni, A. B.,** *Applied Cost Engineering,* Marcel Dekker, New York, 1984.

53. **Brown, D. E.,** Industrial-scale operation of microbial processes, *J. Chem. Technol. Biotechnol.,* 32, 34, 1982.

54. **Sittig, W.,** The present state of fermentation reactors, *J. Chem. Technol. Biotechnol.,* 32, 47, 1982.

55. **Roels, J. A.,** Mathematical models and the design of biochemical reactors, *J. Chem. Technol. Biotechnol.,* 32, 59, 1982.

56. **Anon.,** Cutting down fabrication costs on fermenters, *Brew. Distill. Int.,* 34, June 1982.

57. **Bloch, H. P.,** General type selection factors (sect. 1), Centrifugal and axial turbocompressors (sect. 2), Driver selection (sect. 9), and James, R., Calculation procedures (sect. 8) under Compressors, in *Encyclopedia of Chemical Processing and Design,* Vol. 10, McKetta, J. J. and Cunningham, W. A., Eds., Marcel Dekker, New York, 1979, 157.

58. **Atkinson, B. and Sainter, P.,** Development of downstream processing, *J. Chem. Technol. Biotechnol.,* 32, 100, 1982.

59. **Howell, J.,** Downstream processing, *Pro Bio Tech (suppl. to Proc. Biochem.),* 20, iv, 1985.

60. **Atkinson, B. and Sainter, P.,** Technological Forecasting for Downstream Processing in Biotechnology. Phase 2: Process and Unit Operation Needs, FAST ser. no. 7, Commission of the European Communities, Brussels, 1983.

61. **Bailey, J. E. and Ollis, D. F.,** *Biochemical Engineering Fundamentals,* McGraw-Hill, New York, 1977, 586.

62. **Erikson, R. A.,** Disk stack centrifuges in biotechnology, *Chem. Eng. Prog.,* 80, 51, 1984.

63. **Erikson, R. A.,** personal communication, August 26, 1985.

64. **Hetherington, P. J., Follows, M., Dunnill, P., and Lilly, M. D.,** Release of protein from bakers' yeast (*S. cerevisiae*) by disruption in an industrial homogeniser, *Trans. Inst. Chem. Eng.,* 49, 142, 1971.

65. **Follows, M., Hetherington, P. J., Dunnill, P., and Lilly, M. D.,** Release of enzymes from bakers' yeast by disruption in an industrial homogenizer, *Biotechnol. Bioeng.,* 13, 549, 1971.

66. **Miller, S. A. et al.,** Liquid-solid systems, in *Perry's Chemical Engineers' Handbook,* 6th ed., Perry, R. H., Green, D. W., and Maloney, J. O., Eds., McGraw-Hill, New York, 1984, sect. 19.

67. **Peters, M. S. and Timmerhaus, K. D.,** *Plant Design and Economics for Chemical Engineers,* 3rd ed., McGraw-Hill, New York, 1980, 575.

68. **Boss, F. C.,** Filtration, in *Fermentation and Biochemical Engineering Handbook,* Vogel, H. C., Ed., Noyes Publications, Park Ridge, N.J., 1983, chap. 5.

69. **Rubin, F. L. et al.,** Heat-transfer equipment, in *Perry's Chemical Engineers' Handbook,* 6th ed., Perry, R. H., Green, D. W., and Maloney, J. D., Eds., McGraw-Hill, New York, 1984, sect. 11.
70. **Mehra, D. K.,** Selecting evaporators, *Chem. Eng.,* 93, 56, 1986.
71. **Renshaw, T. A., Sapakie, S. F., and Hanson, M. C.,** Concentration economics in the food industry, *Chem. Eng. Prog.,* 78, 33, 1982.
72. **Strathmann, H.,** Membranes and membrane processes in biotechnology, *Trends Biotechnol.,* 3, 112, 1985.
73. **Lasky, M. and Grant, D.,** Use of microporous hollow fiber membranes in cell harvesting, *Am. Biotechnol. Lab.,* 3, 16, 1985.
74. **Hedges, R. M. and Pepper, D.,** Reverse osmosis concentration and ultrafiltration separation applications in biochemical processes, in *Biotech '83 — Proc. Int. Conf. on the Commercial Applications and Implications of Biotechnology,* Online Publications, Northwood, U.K., 1983, 931.
75. **Cooper, A. R.,** Ultrafiltration, *Chem. Britain,* p. 814, September 1984.
76. **Tutunjian, R. S.,** Scale-up considerations for membrane processes, *Biotechnology,* 3, 615, 1985.
77. **Kristensen, S.,** Application of thin channel membrane filtration in the biotechnological industry, *Pro Bio Tech,* (suppl. to *Proc. Biochem.),* 20, vi. 1985.
78. **Hoare, M. and Dunnill, P.,** Precipitation of food proteins and their recovery by centrifuging and ultrafiltration, *J. Chem. Technol. Biotechnol.,* 34B, 199, 1984.
79. **McGregor, W. C., Ed.,** *Membrane Separations in Biotechnology,* Bioprocess Technologies Series, Vol. 1, Marcel Dekker, New York, 1986.
80. **Curling, J. M. and Cooney, J. M.,** Operation of large scale gel filtration and ion-exchange systems, *J. Parenteral Sci. Technol.,* 36, 59, 1982.
81. **Dwyer, J. L.,** Scaling up bio-product separation with high performance liquid chromatography, *Biotechnology,* 2, 957, 1984.
82. **Anon.,** *Chem. Eng. News,* p. 33, June 24, 1985.
83. **Cazes, J.,** personal communication, December 1985.
84. **Skea, W., Dwyer, J., Kierstead, T., and Findeisen, C.,** Scaling column techniques: simulated process modeling, *J. Anal. Purif.,* 1, 50, 1986.
85. **Reed, P. B.,** Electrodialysis for the purification of protein solutions, *Chem. Eng. Prog.,* 80, 47, 1984.
86. **Jain, S. M. and Reed, P. B.,** Electrodialysis, in *Comprehensive Biotechnology,* Vol. 2, Moo-Young, M., Ed., Pergamon Press, New York, 1985, sect. 37.
87. **LeRoith, D., Shiloach, J., and Leahy, T. J., Eds.,** *Purification of Fermentation Products: Applications to Large-Scale Processes,* American Chemical Society, Washington, D.C., 1985.
88. **Wildfeuer, M. E.,** Approaches to Cephalosporin C purification from fermentation broth, in *Purification of Fermentation Products: Applications to Large-Scale Processes,* LeRoith, D., Shiloach, J., and Leahy, T. J., Eds., American Chemical Society, Washington, D.C., 1985, 155.
89. **Pearson, W. K.,** Advanced batch pilot plant for unloading and drying system, *Chem. Eng. Prog.,* 82, 25, 1986.
90. **Haugen, G.,** personal communication, March 25, 1986.
91. **Perry, R. H., Green, D. W., and Maloney, J. O., Eds.,** *Perry's Chemical Engineers' Handbook,* 6th ed., McGraw-Hill, New York, 1984, 20-33.
92. **Perry, R. H., Green, D. W., and Maloney, J. E., Eds.,** *Perry's Chemical Engineers' Handbook,* 6th ed., McGraw-Hill, New York, 1984, 20-45.
93. **Forrest, J. C.,** Drying processes, in *Biochemical and Biological Engineering Science,* Vol. 2, Blakebrough, N., Ed., Academic Press, London, 1968, chap. 14.
94. **Bailey, J. E. and Ollis, D. F.,** *Biochemical Engineering Fundamentals,* McGraw-Hill, New York, 1977, 334.
95. **Wang, D. I. C., Cooney, C. L., Demain, A. L., Dunnill, P., Humphrey, A. E., and Lilly, M. D.,** *Fermentation and Enzyme Technology,* John Wiley & Sons, New York, 1979, chap. 1 to 6.
96. **Winter, P.,** Process flowsheeting, *Proc. Eng. (England),* 65, 29, 1984.
97. **Okos, M. R. and Reklaitis, G. V.,** Computer-aided design and operation of food processes in industry and academia, *Food Technol.,* 39, 107, 1985.
98. **Felder, R. M., McLeod, G. B., and Moldin, R. F.,** Simulation for the capacity planning of specialty chemicals production, *Chem. Eng. Prog.,* 18, 41, 1985.
99. **Enyedy, G.,** Chemical engineering and cost engineering are getting married finally, *1984 AACE Trans.,* p. B.8.1, 1984.
100. **Enyedy, G.,** personal communication, April 24, 1986.
101. **Perry, R. H., Green, D. W., and Maloney, J. O., Eds.,** *Perry's Chemical Engineers' Handbook,* 6th ed., McGraw-Hill, New York, 1984.
102. **Okos, M. R. and Reklaitis, G. V.,** Computer-aided design and operation of food processes in industry and academia, *Food Technol.,* 39, 107, 1985.
103. **Woods, D. R.,** *Financial Decision Making in the Process Industry,* Prentice-Hall, Englewood Cliffs, N.J., 1975, 291.

104. **Holland, F. A., Watson, F. A., and Wilkinson, J. K.,** *Introduction to Process Economics,* John Wiley & Sons, London, 1974, 117.

105. **Klumpar, I. V. and Slavsky, S. T.,** Updated cost factors. I. Process equipment, *Chem. Eng.,* 92, 73, 1985; II. Commodity materials, *Chem. Eng.,* 92, 76, 1985; III. Installation labor, *Chem. Eng.,* 92, 85, 1985.

106. **Woods, D. R.,** *Financial Decision Making in the Process Industry,* Prentice-Hall, Englewood Cliffs, N.J., 1975, 304.

107. **Fong, W.,** *Fermentation Processes,* Rep. No. 95, Process Economics Program, Stanford Research Institute, Menlo Park, Calif., 1975, 35.

108. **Weaver, J. B., Bauman, H. C., and Heneghan, B. S.,** Cost and profitability estimation, in *Chemical Engineers' Handbook,* 4th ed., Perry, R. H., Chilton, C. H., and Kirkpatrick, S. D., Eds., McGraw-Hill, New York, 1963, sect. 26.

109. **Peters, M. S. and Timmerhause, K. D.,** *Plant Design and Economics for Chemical Engineers,* 3rd ed., McGraw-Hill, New York, 1980, chap. 5.

110. **Black, J. H.,** Operating-cost estimation, in *Cost and Optimization Engineering,* Jelen, F. C. and Black, J. H., Eds., McGraw-Hill, New York, 1983, chap. 15.

111. **Holland, F. A., Watson, F. A., and Wilkinson, J. K.,** Process economics, in *Perry's Chemical Engineers' Handbook,* 6th ed., Perry, R. H., Green, D. W., and Maloney, J. O., Eds., McGraw-Hill, New York, 1984, sect. 25.

112. **Hickman, W. E. and Moore, W. D.,** Managing the maintenance dollar, *Chem. Eng.,* 93, 68, 1986.

113. **Gray, R. J. and Pesek, V.,** Petroleum-coke-fired cogeneration, *Chem. Eng. Prog.,* 81, 70, 1985.

114. **Berthrong, P., Goldberg, R., Finnerty, W., Karlan, D., and Mansfield, P.,** Managing the pharmaceutical QC laboratory with LIMS, *Am. Biotechnol. Lab.,* 4, 20, 1986.

115. **Bruck, J. M., Gurney, D. P., and Jordan, R. H.,** Examples of Economic and Engineering Concepts in the Treatment of Fermentation Effluents, presented at AIChE 1983 Annual Meeting, Washington, D.C., October 30 to November 4, 1983.

116. Technical and Economic Assessment of Processes for the Production of Butanol and Acetone, NASA-CR-169623 (DOE/ECUT), Chem Systems, Inc., Tarrytown, N.Y., September 1982.

117. **Lenz, T. G. and Moreira, A. R.,** Economic evaluation of the acetone-butanol fermentation, *Ind. Eng. Chem. Prod. Res. Dev.,* 19, 478, 1980.

118. **Marlatt, J. A. and Datta, R.,** Acetone-butanol fermentation process development and economic evaluation, *Biotechnol. Prog.,* 2, 23, 1986.

119. **Moo-Young, M., Daugulis, A. J., Chahal, D. S., and Macdonald, D. G.,** The Waterloo process for SCP production from waste biomass, *Proc. Biochem.,* 14, 38, 1979.

120. **Moo-Young, M., Ling, A., and Macdonald, D. G.,** New process changes pulpmill waste to animal food, *Pulp Pap. Can.,* 81, T125, 1980.

121. **Moo-Young, M., Macdonald, D. G., and Ling, A.,** Sensitivity analysis of the Waterloo SCP process in agricultural waste treatment, *Can. Agric. Eng.,* 22, 119, 1980.

122. **Cross, E. B.,** personal communication, June 25, 1985.

123. **Abbott, B. J. and Clamen, A.,** The relationship of substrate, growth rate, and maintenance coefficient to single cell protein production, *Biotechnol. Bioeng.,* 15, 117, 1973.

124. **Moo-Young, M.,** Economics of single cell protein production, *Proc. Biochem.,* 12, 6, 1977.

125. **Fong, W. S., Jones, J. L., and Semrau, K. T.,** Costs of producing ethanol from biomass, *Chem. Eng. Prog.,* 76, 39, 1980.

126. **Keim, C. R.,** Technology and economics of fermentation alcohol — an update, *Enzyme Microb. Technol.,* 5, 103, 1983.

127. **Oliver, S. G.,** Economic possibilities for fuel alcohol, *Chem. Ind.,* p. 425, June 18, 1984.

128. **Bowman, L. and Geiger, E.,** Optimization of fermentation conditions for alcohol production, *Biotechnol. Bioeng.,* 26, 1492, 1984.

129. **Esser, K. and Karsch, T.,** Bacterial ethanol production: advantages and disadvantages, *Proc. Biochem.,* 19, 116, 1984.

130. **Hartley, B. S., Payton, M. A., Pyle, D. L., and Shama, G.,** Development and economics of a novel thermophilic ethanol fermentation, *Biotech '83 — Proc. Int. Conf. on the Commercial Applications and Implications of Biotechnology,* Online Publications, Northwood, U.K., 1983, 895.

131. **Tegtmeier, U.,** Process design for energy saving ethanol production, *Biotechnol. Lett.,* 7, 129, 1985.

132. **Hamer, G.,** Impacts of economic strategies on biotechnological developments, *Trends Biotechnol.,* 3, 73, 1985.

133. **Busche, R. M.,** The business of biomass, *Biotechnol. Prog.,* 1, 165, 1985.

134. **Clausen, E. C. and Gaddy, J. L.,** Organic acids from biomass by continuous fermentation, *Chem. Eng. Prog.,* 80, 59, 1984.

135. **Swartz, R. W.,** The use of economic analysis of penicillin G manufacturing costs in establishing priorities for fermentation process improvement, in *Annual Reports on Fermentation Processes,* Vol. 3, Perlman, D. and Tsao, G. T., Eds., Academic Press, New York, 1979, chap. 4.

136. **Black, J. H.,** Operating cost estimation, in *Cost and Optimization Engineering,* Jelen, F. C. and Black, J. H., Eds., McGraw-Hill, New York, 1983, chap. 15.

137. **Harrison, F. G. and Gibson, E. D.,** Approaches for reducing the manufacturing costs of 6-aminopenicillanic acid, *Proc. Biochem.,* 19, 33, 1984.

138. **Sahai, O. and Knuth, M.,** Commercializing plant tissue culture processes: economics, problems and prospects, *Biotechnol. Prog.,* 1, 1, 1985.

139. **Fuller, K. W.,** Chemicals from plant cell cultures — some biochemical and physiological pointers, *Chem. Ind.,* p. 825, December 3, 1984.

140. **Anon.,** *Biotechnology,* p. 120, February 1985.

141. **Anon.,** Industry outlooks, *Business Week,* p. 248, March 22, 1985.

142. **Clausi, A. S.,** Interfaces of the food industry with biotechnology, *Food Drug Cosmet. Law J.,* 40, 259, 1985.

143. **Korwek, E. L.,** FDA regulation of biotechnology as a new method of manufacture, *Food Drug Cosmet. Law J.,* 37, 289, 1982.

144. **Baum, R. M.,** Environmental policy issues still plague biotechnology research, *Chem. Eng. News,* p. 33, June 17, 1984.

145. Food and Drug Administration, A Plan for Action, U.S. Public Health Service, Department of Human Health Services, p. 44, July 1985.

146. Office of Science and Technology Policy (OSTP), Proposal for A Coordinated Framework for Regulation of Biotechnology, Fed. Reg. 49, 50856, 1984.

147. **Middlekauff, R. D. and McCue, W. A.,** Regulation of the products of biotechnology in the food industry, *Food Technol.,* 39, 73, 1985.

148. **Jones, D. D.,** How the federal government will oversee food products of emerging biotechnologies, *Food Technol.,* 39, 59, 1985.

149. **Jones, D. D.,** Commercialization of gene transfer in food organisms: a science-based regulatory model, *Food Drug Cosmet. Law J.,* 40, 477, 1985.

150. **Korwek, E. L. and de la Cruz, P. L.,** Federal regulation of environmental releases of genetically manipulated microorganisms, *Rutgers Comput. Technol. Law J.,* 11, 301, 1985.

151. **Anon.,** Biotechnology products: EPA finds gaps in risk assessment, *Chem. Eng. News,* p. 6, February 17, 1986.

152. **Gibbs, J. N. and Kahan, J. S.,** Biotechnology and the food industry: leaping the regulatory hurdles, *Biotechnology,* 4, 199, 1986.

153. **Daly, P.,** *The Biotechnology Business,* F. Pinter, London, 1985.

154. **Korwek, E. L.,** personal communication, December 31, 1985.

155. **O'Sullivan, D.,** OECD set to issue DNA safety guidelines, *Chem. Eng. News,* p. 22, February 10, 1986.

156. **Knudsen, I.,** Potential food safety problems in genetic engineering, *Reg. Toxicol. Pharmacol.,* 5, 405, 1985.

157. **Harrison, F. G.,** Current good manufacturing practices for biotechnology-oriented companies, *Biotechnology,* 3, 42, 1985.

158. **Casella, P. F. and D'Agostino, R. A.,** Patenting microorganisms and the role of the International Depository Authority, *Am. Biotechnol. Lab.,* p. 26, July/August 1985.

159. **Beier, F. K., Crespi, R. S., and Straus, J.,** *Biotechnology and Patent Protection, An International Review,* Organization for Economic Co-operation and Development, Paris, 1985.

160. **Benson, R. H.,** Biotechnology patent pitfalls, *Biotechnology,* 4, 118, 1986.

161. **Lawrence, R.,** Procedures and pitfalls in patent protection, *Biotech '83 — Proc. Int. Conf. on the Commercial Applications and Implications of Biotechnology,* Online Publications, Northwood, U.K., 1983, 121.

162. **Hofer, M. A.,** U.S. biotechnology considerations — a corporate view, *Biotech '85 Vol. 2: USA,* Online Publications, London, 1985, 17.

163. **Anon.,** *Chem. Week,* p. 40, April 30, 1986.

164. **Ryan, W. J.,** Impact of the act on research-based industry, *Food Drug Cosmet. Law J.,* 40, 345, 1985.

165. **Marcus, I.,** Fermentation processes and products: problems in patenting, in *Microbial Technology,* Vol. 2, Peppler, H. J. and Perlman, D., Eds., Academic Press, New York, 1979, chap. 19.

166. **Wang, D. I. C., Cooney, C. L., Demain, A. L., Dunnill, P., Humphrey, A. E., and Lilly, M. D.,** *Fermentation and Enzyme Technology,* John Wiley & Sons, New York, 1979, 212.

167. **Wang, D. I. C., Cooney, C. L., Demain, A. L., Dunnill, P., Humphrey, A. E., and Lilly, M. D.,** *Fermentation and Enzyme Technology,* John Wiley & Sons, New York, 1979, 194.

168. **Fox, R. I.,** Computers and microprocessors in industrial fermentation, in *Topics in Enzyme and Fermentation Biotechnology 8,* Wiseman, A., Ed., Ellis Horwood, West Sussex, England, 1984, chap. 4.

169. **Buckland, B., Brix, T., Fastert, H., Gbewonyo, K., Hunt, G., and Jain, D.,** Fermentation exhaust gas analysis using mass spectrometry, *Biotechnology,* 3, 982, 1985.

170. **Clarke, D. J., Blake-Coleman, B. C., Calder, M. R., Carr, R. J., and Moody, S. C.,** Sensors for bioreactor monitoring and control — a perspective, *J. Biotechnol.,* 1, 135, 1984.

171. **Turner, A. P.,** Biosensors for process monitoring and control, *The World Biotech Report 1985,* Vol. 1, Online Publications, London, 1985, 181.

172. **Fitts, R.,** Biosensors for biological monitoring, in *Biotechnology in Food Processing,* Harlander, S. K. and Labuza, T. P., Eds., Noyes Publications, Park Ridge, N.J., 1986, chap. 18.

173. **Wang, N. S. and Stephanopoulos, G. N.,** Computer applications to fermentation processes, *CRC Crit. Rev. Biotechnol.,* 2, 1, 1984.

174. **Knorre, W. A., Guthke, R., and Bergter, F.,** On-line control of fermentation processes — an overview, in *Overproduction of Microbial Products (FEMS Symp. No. 13),* Krumphanzl, V., Sikyta, B., and Vanek, Z., Eds., Academic Press, London, 1982, 623.

175. **Meiners, M.,** Computer application in fermentation, in *Overproduction of Microbial Products (FEMS Symp. No. 13),* Krumphanzl, V., Sikyta, B., and Vanek, Z., Eds., Academic Press, London, 1982, 637.

176. **Meiners, M. and Rapmundt, W.,** Some practical aspects of computer applications in a fermentor hall, *Biotechnol. Bioeng.,* 25, 809, 1983.

177. **Spark, L. B.,** Integrated process control for product development and production, *The World Biotech Report 1985,* Vol. 1, Online Publications, London, 1985, 203.

178. **Williams, D., Yousefpour, P., and Swanick, B. H.,** Online adaptive control of a fermentation process, *IEE Proc. (England),* 131, 117, 1984.

179. **Bown, G.,** Microcomputer controlled batch sterilization in the food industry, *Chem. Ind.,* p. 359, June 3, 1985.

180. **Saguy, I., Ed.,** *Computer-Aided Techniques in Food Technology,* Marcel Dekker, New York, 1983, chaps. 10 and 12.

181. **Bozenhardt, H.,** Multilevel integrated batch control, *Chem. Eng. Prog.,* 81, 35, 1985.

182. **Cooney, C. L.,** Application of process control to fermentation processes, in *Biotechnology and Bioprocess Engineering,* Ghose, T. K., Ed., Link House, New Delhi, 1985, 381.

183. **Modelevsky, J. L.,** Computer applications in applied genetic engineering, *Adv. Appl. Microbiol.,* 30, 169, 1984.

184. **McCormick, D.,** Microcomputer tools for biotechnology, *Biotechnology,* 2, 1022, 1984.

185. **Anon.,** New software package, *Genet. Eng. Biotechnol. Monitor (UNIDO),* No. 10, 56, November/December 1984.

186. Guide to Inspection of Computerized Systems in Drug Processing, issued by National Center for Drugs and Biologics and Executive Director of Regional Operations, Public Health Service, Food and Drug Administration, Washington, D.C., February 1983.

187. **Barsamian, J. A.,** Justifying plantwide computer control, *Chem. Eng.,* 93, 105, 1986.

188. **Bailey, J. E. and Ollis, D. F.,** Sterilization reactors, in *Biochemical Engineering Fundamentals,* McGraw-Hill, New York, 1977, 545.

189. **Parker, A.,** Sterilization of equipment, air and media, in *Biochemical Engineering,* Steel, R., Ed., Macmillan, New York, 1958, 95.

190. **Pym, D. P.,** Aseptic design of fermentation systems, *Proc. Biochem.,* p. 37, February 1984.

191. **Gaither, N.,** *Production and Operations Management,* 2nd ed., Dryden Press, New York, 1984, chap. 7.

192. **Guidoboni, G. E.,** Engineering for an economic fermentation, *Chem. Ind.,* p. 439, June 18, 1984.

193. **McEvan, V.,** Project management — key to success for biological plant construction, *Biotech '83 — Proc. Int. Conf. on the Commercial Applications and Implications of Biotechnology,* Online Publications, Northwood, U.K., 1983, 259.

194. **Davis, D.,** New projects: beware of false economies, *Harv. Bus. Rev.,* 63, 95, 1985.

195. **Holland, F. A., Watson, F. A., and Wilkinson, J. K.,** Process economics, in *Perry's Chemical Engineers' Handbook* 6th ed., Perry, R. H., Green, D. W., and Maloney, J. O., Eds., McGraw-Hill, New York, 1984, sect. 25.

196. **Clawson, R. T.,** Controlling the manufacturing start-up, *Harv. Bus. Rev.,* 63, 6, 1985.

197. **Bartholomew, W. H. and Reisman, H. B.,** Economics of fermentation processes, in *Microbial Technology,* Vol. 2, 2nd ed., Peppler, H. J. and Perlman, D., Eds., Academic Press, New York, 1979, chap. 18.

198. **Stowell, J. D. and Bateson, J. B.,** Economic aspects of industrial fermentation, in *Bioactive Microbial Products 2: Development and Production,* Nisbet, L. J. and Winstanley, D. J., Eds., Academic Press, London, 1983, 117.

199. **Black, J. H.,** Cost effectiveness, in *Cost Engineering Planning Techniques for Management,* Marcel Dekker, New York, 1984, chap. 7.

200. **Jelen, F. C.,** Cost engineering and beyond, in *Cost and Optimization Engineering,* Jelen, F. C. and Black, J. H., Eds., McGraw-Hill, New York, 1983, 461.

201. **Holland, F. A., Watson, F. A., and Wilkinson, J. K.,** *Introduction to Process Economics,* John Wiley & Sons, London, 1974, chaps. 1 to 4.
202. **Woods, D. R.,** *Financial Decision Making in the Process Industry,* Prentice-Hall, Englewood Cliffs, N.J., 1975, chaps. 3 to 5.
203. **Aragon, G. A.,** *A Manager's Complete Guide to Financial Techniques,* Free Press, New York, 1982, parts 1, 3, and 5.
204. **Peters, R. A.,** *ROI — Return on Investment,* AMACOM, New York, 1979.
205. **Holland, F. A., Watson, F. A., and Wilkinson, J. K.,** *Introduction to Process Economics,* John Wiley & Sons, London, 1974, appendix 1.
206. **Anon.,** Present value of future revenue at various discount factors, *Pet. Eng. Int.,* 57, 81, 1985.
207. **Powell, T. E.,** A review of recent developments in project evaluation, *Chem. Eng.,* 92, 187, 1985.
208. **Brealey, R. and Myers, S.,** *Principles of Corporate Finance,* McGraw-Hill, New York, 1984, chaps. 2, 3, and 5.
209. **Mateles, R. I.,** Economic analysis of genetic engineering: single-cell protein, *Chem. Eng. Commun.,* 45, 213, 1986.

INDEX

234   *Economic Analysis of Fermentation Processes*

Return on investment (ROI), 5, 187, 197, 210
  depreciation and, 107
  differential, 196
  quantitative measure of, 198
Reverse osmosis, 71, 72
Riboflavin, 22
Risk analysis, 47
Rotary drum filter, 68
Rotary dryer, 84
Rotary vacuum filter, continuous, 68
Rotary vacuum filtration, 18
Royalty costs, 108
RPM control, 151, 152

S

Safety regulations, 32, 143
Sales/marketing costs, 100
Sales plan, 190
Sanitation, 112
Scale-up, 5
  considerations in, 52
  economic analysis of, 7—9
  maintenance of equivalency on, 15
  parameters of, 9
  requirements of, 36
Scraped surface evaporators, 70
Seed vessels, 34—35
Sensitivity analysis, 210
Shutdowns, 40
Single cell proteins (SCP), 19—20
  analysis of, 130—131
  preparation of, 39
  process, 126, 127
Sludge, 117
Solids ejecting centrifuges, 64
Solvents
  extraction, 39
  recovery, 18, 83
  volatile, 72
  vapor, 69
Soybean, 25
Soybean meal, 26
Soy meal, 25
Space requirements, 43—44
Speciality chemical production, 89
Spills, 42, 115
Spiral plate exchangers, 70
Spray drying, 83—84
Standard cost, 210
Standard unit cost, 97
Start-up, 180—186
  commissioning phase of, 185
  completion of, 185—186
  cost estimate for, 184
  costs of, 181—182, 210
  data for, 7—9
  personnel assignments for, 182
  planning in, 183
  preparation for, 181
  steps of, 183—185

  team, 182—183
  training during, 185
Start-up companies, 121, 213—214
  rate of, 200
  research and development costs for, 2—3
Statistical analyses, 162
Steam
  costs of, 102—104
  energy, 102
  low-pressure, 110
  sterilization, 102
  turbines, 102
Sterile design
  check lists for, 172—173
  costs of, 173—174
  goal of, 170
  principles of, 170—172
Sterile filtration, 172
Sterile fluid flow, 153
Sterility, 36, 72, 170—174
  break in, 12—13
  maintenance of, 149
Sterilization, 12
  automation of, 162—163
  considerations in, 13
  continuous, 12—13
  temperature range for, 150
Stirred tank gas-liquid contactor design, 9
Stirred tank reactor, 52
Storage conditions, 14
Storm sewers, 115
Straight-line depreciation, 107
Strain gauge systems, 152
Strain improvement, 14, 21
String discharge filter, 68
Substrates, see Media
Substrate utilization rates, 13
Sucrose, 22, 23
  cost of, 135
Sugar
  costs of, 23—25
  pricing of, 30
  sources of, 22—25
Sulfur, 18—19
Sunk costs, 210
Supernatant, 63
  clear, 71
Supervision, costs of, 106

T

Temperature control, 150
Temperature indication, 13
Thermoanaerobium brockii, 130
Thermophilic organisms, 130
Thin-film evaporators, 70
Tissue plasminogen activator, 200
Tower dryer, 84
Toxic materials, regulatory requirements for, 34
Toxicology testing, 168
Transfer price, 101

Printed in the United States
by Baker & Taylor Publisher Services